Understanding the basics of energy

Understanding the basics of energy

An introduction

from simple to complex situations

Harald Mehling

Impressum

© 2019 Harald Mehling

1. Auflage / 1st edition

Verlag und Druck: tredition GmbH, Halenreie 40-44, 22359 Hamburg

ISBN Paperback: 978-3-7482-0896-9

ISBN Hardcover: -

ISBN e-Book: -

Bibliografische Information der Deutschen Nationalbibliothek:

Die Deutsche Nationalbibliothek verzeichnet diese Publikation in der Deutschen Nationalbibliografie; detaillierte bibliografische Daten sind im Internet über http://dnb.d-nb.de abrufbar.

Contact address:

Harald Mehling, Weingartenstr. 37, 97072 Würzburg, GERMANY,

e-mail: *harald.mehling@gmail.com*,

LinkedIn: *www.linkedin.com/in/harald-mehling*,

ResearchGate: *www.researchgate.net/profile/Harald_Mehling*,

Amazon Authors: *www.amazon.com/author/harald_mehling*

Acknowledgements

No book on science and technology is based on work of the author only. There is always a historic basis. So, besides adding new findings, an author often presents already known things more comprehensive, puts things in a different perspective or structure, or explains things in a different way.

This book is based on the work of many people who have contributed to the field of "energy" in the past decades, actually centuries. They observed what happens, made detailed measurements, discovered effects, developed terms, concepts and laws or rules, then questioned, tested, and revised them. Thus, they have accumulated all the knowledge that this book is based on. The most important ones are famous, and are mentioned in a historic review together with their work. If specific information is given they are cited. However, for many things in science the original contributor is unknown. Therefore, I would like to thank all of them here.

Besides having myself some open questions from writing the last book, there has been a second, strong source of motivation to write this new book. This source is the report of a workshop "Teaching about energy", organized by Paula R.L. Heron and Marisa Michelini. I would like to thank the authors of the report and all those who contributed. Because I have never been teaching on the school level it was a motivation, and also an overview of the crucial questions, problems, and therefore a guide to select the content of this book and the examples and explanations used.

I would also like to thank David Watson, who runs a website on energy basics, and who was willing to discuss many issues regarding what energy is and other related and important questions, possible answers, supply information etc. This book would never have developed this far without him.

Finally, I would like to thank Joan Roca and Eva Günther for reviewing parts of the book, and valuable comments on its content and structure.

Harald Mehling

I

Preface

Energy is a fascinating topic. Energy shows up in so different things like the fall of an apple, the swing of a pendulum, the motion of the air in the wind, the sunlight, or the use of fuels in chemical reactions. It is amazing that a single term covers so many different things. Energy relates changes in very different things like velocity, height, amount of fuel, and it allows understanding and predicting these changes, even affecting and optimizing them. This is central to science and engineering, moreover to our daily life!

While energy with respect to our energy system is obviously important, most people are unaware of its significance in the supply of materials, food, and with increasing importance now information. Energy is crucial in our lives anywhere and at any time. Yet, most people know only little about it.

Literature on the scientific and technical aspects of energy is abundant, however usually at an advanced level, written for people already having a scientific or technical degree, or being in a later stage of education towards such a degree. Literature at an educational level for the first semester at university or below is rare, often vague or even erroneous, and thus unsuitable. Therefore, I decided to write about energy related matters on a level useful for students and their professors for introductory courses, and for teachers to prepare their classes. Thus, basic high school physics and chemistry should be sufficient for reading these books. The books should give an overview of their topics, containing a systematic, clear, and comprehensive treatment. And they should explain why things are the way they are, explain common misconceptions, such that at the end the reader really understands things.

With the book "Technologies of energy conversion, storage, and transport in the energy system – A brief introduction", published in 2016, I covered the energy technologies, meaning the engineering side. The book describes the different technologies and how they work together to finally form the energy system. The book is a good introduction and overview, even for people having a scientific or technical degree, or studying towards such a degree. The basics of energy are treated in a way covering what is essential to understand how the different technologies work, and for a summary and reference. Thus, the discussion of the basics of energy is only to a depth necessary to understand the technologies.

While writing the respective chapter I realized that the basics of energy are a topic for an own book. I realized that energy forms are often classified in different ways; definitions are frequently only valid in special cases, wrong, or without explanation. In addition, common usage and scientific-technical usage of terms is often different and hence a basis for misunderstandings. Thus, the different terms and concepts are often confusing or not fully clear, even for people working in the field of energy. One reason is that the terms and concepts for the basics of energy have been changed, modified, adapted, extended many times in history. And while we can observe energy associated with the motion and position of a macroscopic object directly, energy in materials is hard to observe and thus to understand. Consequently, our common experience misses crucial things, which in turn leads to energy being mysterious. The following are just a few examples. Crucial is that energy is conserved, but from direct observations we say that it is consumed. We say heat and cold, or heating and cooling, but in science there is only the term heat, and it refers to energy exchanged between systems while for a system the term thermal energy is used. Also, we relate an electric current commonly with electric energy, while in science and technology it relates to electric power. And in addition, power must not be mixed up with force. Most striking is that we do not even have a useful explanation of energy. The most commonly used definition of energy is "the ability to do work", which is as Aristotle initially introduced it. Sometimes "and to supply heat" is added, following the 1st law of thermodynamics "the change of the internal energy of a system is equal to the amount of heat supplied to the system, and less the work done by the system", which was stated much later. Another definition in literature is "Work is the energy used to cause an object to move against a force". So we have statements like "work is energy ...", "energy is the ability to do work", and "changes of energy are equal to the amount of work ...". Not only does none of them explain what energy is, they do not even agree on the relation between energy and work. And in contrast to common thinking, even when just looking at practical problems thermodynamics does not give a comprehensive description of energy; thermodynamics is restricted to internal energy, usually referring to materials and thus missing macroscopic objects. The problems for understanding the basics of energy are thus quite fundamental. The report of the workshop "Teaching about energy", organized by Paula R.L. Heron and Marisa Michelini, finally convinced me to write a book specifically on the basics of energy, and in many ways, this report also acted as a guide on the content.

III

Despite these problems, the fact that "energy" is used successfully for a long time indicates that there should also be a way to understand its basics. With 25 years of experience in energy research and also material properties, I was confident that I already understand the "basics of energy" somehow. But knowing to apply something and teaching its application is one thing. To understand the basics of energy, beyond just its application, and to write a systematic, clear, and comprehensive treatment of its basics is something very different. This is probably why the topic has been avoided for so long.

The key question is: what is energy? Connected to this, because of its practical relevance, is that energy is related to changes. It appears in different energy forms, and energy can be converted between them. As a condition, science tells us that the total amount of energy is conserved, and conversion is limited in some way. These are the key issues to be addressed.

Now how is it possible to understand them? The widespread use of the term "energy" started while there was still little known on what is going on in materials. Thus, initially the effects related to energy were not understood. Today, it is common knowledge that materials are composed of particles, e.g. atoms, and that the observed effects can be understood and explained from the motion and interaction of them. This is the key approach used in this book. Models for specific effects related to energy are plenty and well described. But a single model for a specific effect gives limited insight, not a general picture. And this is also not changing when having many of them. But generally, it is clear that the basics of energy can be understood looking at the motion and interaction of objects, from simple to complex situations.

For an overview of the basics of energy, a systematic, clear, and comprehensive treatment, it is necessary to start from simple situations and then widen the view to complex situations. And this not only for a few specific effects. The simple situations are the motion of one object and its interaction with a second one. From this, complex situations with many objects and different types of interactions can be understood. And from them follow all energy forms, and conversion, storage, and transport processes. Starting with effects of one and two objects allows to stay simple, refer to every-day experience, in fact follow in many ways the historic development and explain where in history misconceptions were discovered, removed, and the terms and concepts adapted.

IV

As an introduction and overview on the basics of energy, the text covers the terms and concepts of force, momentum and impulse, energy, specifically the different energy forms and conversion between them, energy conservation, and heat and work. It starts from simple situations, which are one object in motion or interacting with a second one, and goes all the way to the concept of energy using energy forms to describe the energy of systems, energy exchange between systems, and energy conservation and conversion.

For a real understanding, examples relating things to common experience, simulations and experiments, movies etc. are useful, especially for teaching. Some are given in the text. It is also planned to make more available via the internet, but at the time of publication no details were decided yet.

After two years working on this book, the final text has developed a lot from the initial idea and key questions, to the final content and the structure. Selecting material and finding suitable ways to present and explain things is crucial. Even more, trying to find the logic and consistent structure behind things is a difficult task. Many things had to be reconsidered and rewritten, such that crafting the text has been quite a learning process, and in several ways even a discovery of new things. Looking back now, this is no surprise. This book is the first attempt to cover the basics of energy in a systematic and comprehensive way.

Despite that a systematic treatment is a way to discover inconsistencies, there is always a chance that a text contains other errors or inaccuracies. Further on, there is always potential to improve things, e.g. explanations or examples. Thus, any feedback to improve this book is highly appreciated.

I hope the reader enjoys reading this book as much as I enjoyed writing it, and finds it a valuable introduction to the fascinating topic of energy.

Harald Mehling

Würzburg, 2019

Contents

1 Scope, road map, and fundamentals

Chapter 1 gives the scope of the book, a road map of its content, and repeats fundamental terms and concepts. Let's start with the historic development of science regarding things related to energy; it will help to understand many things later. The following historical review is based on Rooney 2011, with additional information on energy from Snurr and Freude 2014.

The origin of science is natural philosophy; people, called philosophers, discussed nature, especially what things are made of, their origin, changes, and the reasons for everything of course. Our fundamental concepts are that **space and time refer to where and when things happen, matter refers to what the things like macroscopic objects are made of, and force and energy relate to why and how things change**. Early philosophers however used supernatural explanations, usually one or several gods. With the advent of the scientific method for investigation and thus science, people tried to identify natural reasons for what they observed, develop reliable rules and later mathematical descriptions, instead of using supernatural explanations. As a result, the understanding of the reasons behind the observed changes increased, people learned to describe their observations mathematically, to apply the knowledge to their advantage, and philosophers following this approach were from then on called scientists. But being based on observations does not mean that concepts, once developed, remain forever. Even the ideas about such fundamental things as space and time changed in history.

Scientific thinking probably started with Thales of Miletus (624 to 546 BC); he suggested that there is a physical cause for all observed phenomena, meaning related to bodies, instead of a mental, spiritual, supernatural cause. This started the search for the physical causes. About a century later, Anaxagoras (about 500 to about 430 BC) was thinking about the universe as being explicable by the rational mind, thus saying that we are actually able to find the physical causes and understand things. Aristotle (384 to 322 BC) then delivered the needed approach, saying that we can gain understanding, and derive the physical laws, by careful observations and measurements; this was the beginning of the scientific method. Later, many significant additions were made on how to make observations, proof or falsify theories, for example by Galileo Galilei (1564 to 1642).

The idea that **matter** is composed of **particles** started in the 5th century BC
with the work of Leucippus and Democritus (about 462 to about 370 BC).
The particles were called "atomos" (the origin of the term atom), meaning
indivisible. But atomic theory was not widely accepted at that time. A parallel model focusing on the properties of matter was that of elements, introduced by Aristotle (384 to 322 BC), starting with 4. In 1789, A. Lavoisier
published a list of 33 elements identified by their chemical properties; in
1868, D. Mendeleev published a more comprehensive and systematic table.
The final rise of atomic theory started when P. Gassendi (1592 to 1655)
proposed that everything that happens is due to the motion and interaction
of particles following natural laws; this made atoms the basis for the composition as well as properties of matter. In 1661, R. Boyle promoted the idea
in his publication "The skeptical chemist". Following this idea, J. Proust
(1754 to 1826) found the law of definite proportions from chemical experiments carried out 1798 to 1804, somewhat after Lavoisier's publication.
Then, J. Dalton (1766 to 1844) laid the basis of modern atomic theory by
giving a comprehensive theory on atoms and elements, published 1808. He
stated: elements are made of atoms, atoms of a given element are identical,
atoms of one element differ from atoms of every other element, atoms cannot be created or destroyed or divided by chemical processes, and atoms of
one element can combine with atoms of another to make a chemical compound which always contains the same proportion of each element. Most
scientists still did not believe in atomic theory without a real proof. The first
evidence that matter is really made up of small particles was discovered by
R. Brown (1773 to 1858), observing under a microscope the random motion
of small particles in a liquid, today called Brownian motion. In 1877 J. De-
saulx explained the observation by collisions with smaller, invisible particles of the surrounding liquid due to their thermal motion. In 1905, A. Einstein derived a mathematical model explaining Brownian motion; atomic
theory was finally proved. But already at that time, evidence of smaller,
subatomic particles emerged. What scientists called atoms, indivisible in
chemical processes, turned out to be divisible in other processes. Already in
1897, J.J. Thomson had discovered the electron, and in 1904 proposed his
"plum pudding" model of atoms. Then, E. Rutherford proved by scattering
experiments that atoms are largely made of empty space, with a positively
charged nucleus and electrons orbiting around it. The nucleus itself consists
of even smaller particles, called protons and neutrons, and both are composed of again smaller particles called quarks that were discovered in 1968.

Now, why and how do things change? This question is probably even more important than what things are made of, as changes determine our daily life. Initially people could investigate only macroscopic **objects**, and the first question was simply: why does a mass move or change its movement? For a long time, people used for example levers to move large stones; they had knowledge of **forces** and rules about them, but no deeper understanding. First investigations on forces are attributed to Aristotle (384 to 322 BC), based on observations of a balance with lever arms of different length. Later, Archimedes (287 to 212 BC) stated that weight times distance is the same on both sides, thus giving not just a general but a quantitative relation. But much more problematic was dynamic motion. Aristotle proposed that something moves because a force is applied to it, but this fails to explain for example why a stone continues to fly after it is thrown and not any more affected by the person that threw it. One idea was that the person transferred "something" connected with motion to the stone. But then, why does a pendulum continue to swing at its lowest point? The solution was the introduction of **inertia** and **momentum**. Galileo Galilei (1564 to 1642) then made experiments on gravitation; to slow down free fall, he let balls roll down slopes. He observed that light and heavy balls roll the same way and thus concluded that gravity works the same on both. That feathers fall slow he explained by air resistance, and thus an external effect. Making experiments with even shallower slopes, he concluded something else: without a force to stop it, a ball would roll forever; again, he attributed the observation that it does not roll forever to the external effect of friction. The further development of science was significantly based on the development of new tools, mathematical tools for the description and analysis of position and motion, and experimental tools for observation and measurement. Regarding the measurement of **time**, the sun and the moon were useful for timescales of a fraction of a day or larger, but Galilei still had to use his pulse to measure short times until he observed that a pendulum swings in a periodic way; following his observations, C. Huygens built the first pendulum clock in 1656. And Galilei also significantly improved the telescope in 1609, which was previously invented by H. Lippershey, and discovered e.g. the moons of Jupiter and the phases of Venus. As a result, most astronomers started to believe in the heliocentric model saying that the planets move around the sun. Regarding mathematics, Cartesian coordinates (x, y, z) to specify a point in **space** were discovered by R. Descartes (hence the name) in 1619; he watched a fly in his bedroom and realized that its position and movement in three-dimensional space can be described by the distances to the walls and ceiling. And Isaac Newton (1642 to 1727) developed differential calculus. As a result, a much more detailed mathematical analysis of observations and mathematical formulations of laws became possible. Newton then used the new mathematical tools, and based on previous works e.g. of Galilei, published his famous three laws of motion in 1687, and stated the conservation of linear and angular momentum. And he formulated the law of gravitation, which allowed to calculate the motion of planets and the fall of apples, thus proving that physical laws do not only apply on earth but are universal.

What about **energy**? Its original meaning is from the Greek term "energeia", composed of "en", meaning "within" and "ergon", meaning "work". Even today, **"the ability to do work"** is used frequently as **definition of energy**. The term "energeia" was first used by Aristotle (384 to 322 BC), but then not used for a long time. The first observation towards what later becomes the concept of energy is probably by Galilei, who observed some kind of **conversion** between motion and height when a pendulum swings. G.W. von Leibniz (1646 – 1716) then mathematically described the conversion, and by 1686 he had developed concepts that correspond to kinetic and potential energy, however, he did not use the term "energy" but used the term **"vis viva"**. É. du Châlet (1706 to 1749) then described what is today called kinetic energy of a moving body by its mass times the square of its velocity; later the factor ½ was added. Thus, the first ideas behind the concept of energy started to emerge at the end of the 17th century, but the term "energy" in its current use emerged later, in the 19th century, starting in 1807 with T. Young. Earlier observations were limited to the motion or position of macroscopic objects, but at that time it had become clear that energy comes in more different forms, and that not all are macroscopic. For example, boring holes in metal, such as cannon barrels, showed that mechanical work produced heat. James Prescott Joule (1818 to 1889) performed own experiments to show that and made quantitative measurements. While it was common that mechanical work produced heat by friction, the first steam engine developed by D. Papin in 1690 showed that heat can also be used to get work; and new sources for work were required. Improvements were made by T. Newcomen in 1712 and by J. Watt in 1769, and **"work"** was defined in 1826 by G.-G. Coriolis as **"weight lifted through a height"**, based on using steam engines to lift buckets of water out of flooded mines (wikipedia 2017a). If heat can do work, then is heat related to energy the same way as work? In the 18th century, there were two theories on what heat is. One theory was that of caloric, proposed by Lavoisier, stating that **heat** is a form of matter, consisting of atoms of heat, flowing like a fluid; hence also the term "heat flow". It gave the unit of calorie its name. The other theory was based on the movement of particles, and therefore on the idea of Gassendi, and ideas published by D. Bernoulli in his book "Hydrodynamica" in 1738. Already Joule concluded that the motion of atoms is the correct explanation, long before the existence of atoms was proved and widely accepted. In the years 1842-1847 then, Julius Robert von Mayer, Joule, and Hermann Ludwig Ferdinand von Helmholtz stated that **"energy cannot be created or destroyed; it can only be transformed from one form to another"**, meaning **energy conservation** and **energy conversion**, but instead of the word "energy" used terms like "living force" and "tensional force" or "fall-force". Then, 1851 to 1852, it was William Thomson (Lord Kelvin) and William J. M. Rankine who finally introduced the term "energy" to replace the related terms using the word "force" in science. The description of observations was then over time collected and summarized in what is today called thermodynamics, which has at its center three laws. The **1st law of thermodynamics**, formulated by R. Clausius (1882 to 1888) is **"the change of the internal energy of a system is equal to the amount of heat supplied to the system, and less the work done by the system"**. Thus, **heat and work refer to energy exchanged between a system and its environment**. The 1st law is based on the observation of Joule on "conversion" between work and heat, combined with the statement of Mayer, Joule, and Helmholtz that energy is conserved but can be transformed. Later, besides kinetic, gravitational, and thermal energy, more energy forms associated with chemical and nuclear reactions, radiation etc. were added. The concept of energy became complex, and what energy is even less clear.

While there are many energy forms, initially often not understood, the conservation of energy derived from observations was always reliable; it relates changes in one energy form to changes in another, and is often the crucial point for applying energy. Matter also cannot come into existence from nothing or disappear into nothing; already Anaxagoras (about 500 to about 430 BC) thought so, but A. Lavoisier (1743 to 1794) later stated it based on experiments as the **conservation of mass**, saying **"mass is never lost or gained in the process of a chemical reaction"**. Later, Einstein showed with his famous formula $E = m \cdot c^2$ that energy and matter are connected.

1.1 Scope and roadmap

The most commonly used definition of energy is "the ability or capacity to do work", similar to the meaning introduced by Aristotle (384 to 322 BC). It explains the use of energy, but not what it is. Even more, compared to the 1st law of thermodynamics stated by Clausius (1882 to 1888), it misses heat, and also that work and heat conversely lead to a change in energy. Today, after centuries of development of the concept of energy and its successful use in many areas, there is no good explanation of what energy actually is. For an answer, a comprehensive treatment is needed.

Goal

The goal of this book is to give a systematic, clear, and comprehensive treatment of the basics of energy: the concept and the physical basis.

Key issues

The key issues that need to be addressed about energy basics can be summarized in a few statements.

A common way to say what energy is, is "energy is the ability to do work". Work, in a simple way, is a force acting a distance, so energy allows moving something for a distance against a force. Thus, of practical relevance is that **energy allows us to achieve changes**.

In the process of using energy, **energy appears in different forms**, for example kinetic energy, gravitational energy etc. **What are the energy forms and what is their basis?**

To use energy to our advantage, **energy can be converted between the energy forms**. Energy conversion for example connects available energy forms with those needed for use. **How does energy conversion work? What are the mechanisms?**

Energy conversion is limited, first because **the total amount of energy is conserved**, and second because **entropy tends to increase. Why is energy conserved? What is entropy, and why does it tend to increase?**

All combined will explain the concept of energy and its physical basis, thus answer the question "what is energy?"

Stage

We all have common, fundamental concepts about the world around us. **Space refers to where things are, and time is used to describe when, specifically to describe processes of change. Things we are familiar with are objects, like a stone, a person, the moon … Matter refers to what the things are made of; specifically large objects consist of smaller ones, commonly called particles. Forces relate to why and how things change, specifically the position of objects (including small particles) by motion (change of position). The forces are due to interactions between objects.** These concepts come from everyday life; everybody is familiar with them, even children, maybe except particles being so small that they are invisible. Being from everyday life, these concepts also were the basis of investigations back in history. Thus, these concepts form the stage of knowledge in most parts of the book.

In addition to these concepts, science started to describe things by laws. Many things we observe happen reliably, e.g. things fall to the ground. **Laws just state what is observed reliably, usually in many experiments under controlled conditions. The laws then allow everybody to make predictions, which is why they are so useful.** For example, the force of gravitation "explains" that objects fall to the ground reliably, and the law of gravitation allows calculating the force. Common experience usually is about objects with mass; the laws concerning their motion and interaction using the concept of force are called the laws of **classical mechanics**. The term "mechanics" is common in science and technology. Its origin is in ancient Greek, related to "mekhane" meaning "that which enables", like means, tool, device, or machine. As the laws are based on observations, new observations can show that the previous laws are not always accurate, thus lead to modifications; e.g. **quantum mechanics** replaced classical mechanics at small scales, and **relativistic mechanics** replaced it at high velocities. Energy, like forces, relates to why and how things change. But the concept of energy with the law of energy conservation was developed much later than classical mechanics, which uses the concept of forces. The concept of energy became successful by describing things that were initially not understood, e.g. the exchange of heat and work by thermal engines. For this, the laws of **thermodynamics** were developed. Later, often microscopic models were found within classical mechanics that explain things within the range of common experience; they are used here extensively to understand things.

But under certain conditions things cannot be treated within the above stage. Some things are too different from common experience and its concepts. Electromagnetic waves, described by **electrodynamics**, travel without a medium. And by $E = m \cdot c^2$ energy and mass are related and not always conserved individually (e.g. by electron - positron annihilation or creation). Such situations are addressed in the discussion, and finally at the end.

Key ideas – development of a strategy for this book

The concept of energy has today a high practical relevance. What makes energy so useful, but also so complicated, is that it relates so many things, often things that we do not understand from common experience, like heat. The initial statement of Aristotle, as "work within", or its modern form as "energy is the ability or capacity to do work" just expresses the lack of a physical explanation. It is thus maybe not surprising that from its first statement by Aristotle to its final rise in science about 2000 years passed.

To deal with simple situations, specifically with few macroscopic objects, using forces is sufficient, e.g. for Newton to calculate the motion of planets. "Energy" came in use when people made observations that they did not understand, but where the discovery of energy conservation allowed to relate changes in different things, called energy conversion. Then "energy" became useful, and its concept was extended and refined. Observations that cannot be understood directly from what is seen macroscopically relate to materials, where the relevant effects have a microscopic origin and involve an uncountable number of particles. Thus, the concept of energy came in use when dealing with complex situations.

Macroscopic observations of a system with many microscopic particles are the topic of thermodynamics. It started to develop before atomic theory was proved, but particle models later explained many things. Thermodynamics is divided into two fields: macroscopic and microscopic thermodynamics. The microscopic models explain the macroscopic effects, e.g. thermal energy is explained by the microscopic motion and interaction of atoms and molecules. And microscopic models also exist for electric and magnetic effects, for chemical reactions, for deformation etc. The microscopic models can together explain why energy conservation is observed macroscopically, at least in a wide range of cases. However, because thermodynamics does not treat macroscopic objects, it gives no comprehensive account of energy.

The following are the **key ideas** used to start the discussion in this book; they are revised, meaning updated and extended, in section 2.1.7.

For the general strategy:

1. Macroscopic observations of energy conservation or conversion on systems with many objects / particles, like materials, are understood from their microscopic origin.

2. For a comprehensive treatment, microscopic models and macroscopic observations must be discussed together in a single text, and extended beyond just internal energy in materials.

For the structure and roadmap of the discussion:

3. The origin of energy conservation and conversion is in simple situations, which is 1 object in motion, and cases of interaction with a 2nd object. Complex situations are just a combination of many simple situations. Thus, understanding of energy starts with simple situations, and complexity must be increased stepwise to more objects / particles and more interactions to keep track of the origins of effects.

4. To be able to "understand" things it is necessary to stay within the validity of classical mechanics, specifically Newton's laws, as long as possible; this also means including microscopic situations. Newton's laws can still be understood from everyday experience. In general, it is important for any introduction that to understand something an explanation must be based fully on common experience.

For the discussion, especially explanations:

5. In addition, a discussion of the historic background regarding general ideas and use of words is also crucial in an introductory text. It is common to present solutions to a problem, and then try to find a way to explain things. But the best explanation is the historical process of a discovery as it follows how people with common experience think. In addition, we still use terms from outdated concepts like "heat flow", and the meaning of technical terms is not always the same as their colloquial meaning, for example the term "heat". And everyday experience is often limited and incorrect, e.g. when saying "energy consumption". Such cases are misleading and confusing, and therefore they must be addressed and also explained.

Road map

The general structure of the discussion follows the idea that things are best understood from the simple to the complex.

The 1st chapter, SCOPE, ROAD MAP, AND FUNDAMENTALS, introduces the scope of the book, gives its roadmap, and repeats some fundamental terms and concepts that are needed for the discussion in the book.

The 2nd chapter, SIMPLE SITUATIONS WITH 1 OR 2 OBJECTS, starts with an introduction to basic terms, concepts, and laws regarding energy. These are forces, motion, and Newton's laws, then linear momentum, its conservation, and impulse, and finally energy, its conservation, and work. For energy, the terms energy conversion, energy contributions and energy forms are then discussed in more detail, and a preliminary answer to what energy is is given. Afterwards, the fundamental energy contributions, as there are kinetic energy, gravitational energy, electromagnetic energy, and "nuclear energy" are discussed. All energy is based on them.

The 3rd chapter, COMPLEX SITUATIONS WITH MANY OBJECTS, starts with discussing situations with many objects but still a single fundamental energy contribution, thus again kinetic energy, gravitational energy, electric energy, magnetic energy, energy of electric, magnetic and electromagnetic fields, and "nuclear energy". Afterwards energy related to a change in composition is discussed. For this, the atomic model and quantum mechanics are introduced. They lead to excitation energy, ionization energy, chemical energy, and nuclear energy. Finally comes energy related to the relative motion and position of microscopic particles. For this, materials are introduced. This leads to deformation energy and thermal energy.

The 4th chapter, ANSWER TO WHAT ENERGY IS, gives general answers. It starts with a summary, and an analysis of limits of the findings with respect to the stage. It then gives a general answer to what energy is.

The 5th chapter, COMPLETE CONCEPT OF ENERGY, focuses on applying the concept of energy. It briefly repeats the physical basis and general issues, and then the "toolbox" that is needed for calculations, looking at energy of systems, energy exchange between systems, and energy conservation and conversion. The toolbox is then discussed for the concept of energy using fundamental energy contributions, and the concept of energy using energy forms. The chapter ends with a final summary.

1.2 Fundamental terms and concepts

Before starting, it is essential to repeat the fundamental terms and concepts, those that set up the stage for most part of this book.

For most part of history, before the development of tools for observations, people were dealing with objects of a size that is common in their daily life, like a stone (Fig. 1-1), a horse, a house, and similar other things. In addition, people were aware of the earth, its moon, the sun, and some other planets. The invention of the telescope then extended the visible stage, allowing Galilei in 1610 to see for the first time moons of another planet, Jupiter.

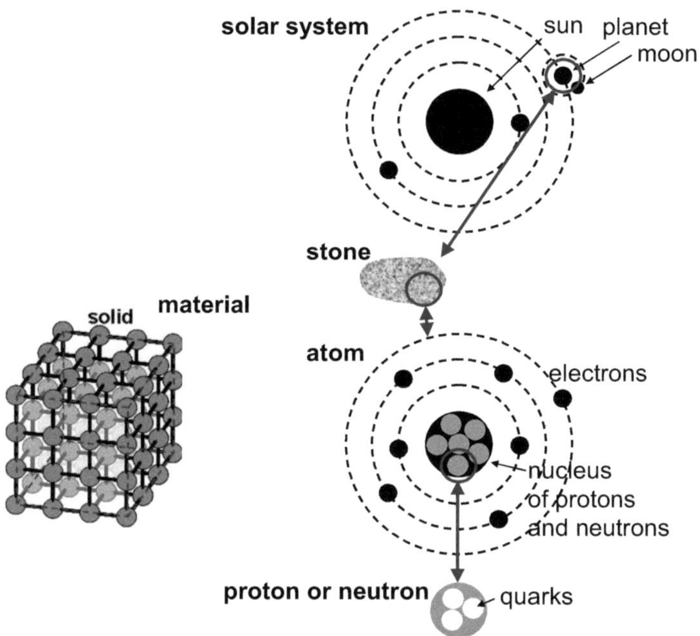

Fig. 1-1 Range of size of objects: objects of common life like a stone, large objects like the planets, and small objects like atoms

The extension to the smaller began with the invention of the microscope in about 1620, and Hook's investigations with it that he published in 1667. Matter, e.g. a stone, is composed of a very large number of atoms (Fig. 1-1). An atom in turn consists of a nucleus, and a number of electrons. And the

11

nucleus consists of protons and neutrons, which consist of quarks, discovered in 1968. Quarks, protons, neutrons, electrons, atoms, are all very small objects, and usually they are a part of larger ones; thus, they are usually called particles. The term **object** is the general term. A **particle** (small part) is a small object. An object can be a single particle or consist of many.

Objects are at the center of the discussion, specifically their position and their motion in **space**, and changes of position and motion with **time**.

The **position** of an object comprises its **location**, meaning where it is, and its **orientation**. The term position is here used as the general one, as it is also used as "to be in the position for doing / achieving … something" in a general manner, while location also just means place. For example, I stand in front of the town hall (location), facing the market place (orientation). The location can also be given with respect to a **reference frame**, e.g. by Cartesian coordinates written as (x, y, z), or by the distance and direction from the point of origin (Fig. 1-2, left). The latter is symbolized by an arrow, called a vector. A **vector** in general has a magnitude (quantity) and a direction, in contrast to a **scalar**, which has only a value. For example, the location is 100 m to the north from the town hall. The result is the same; both are usually called vector. An orientation is given similarly as (α, β, γ).

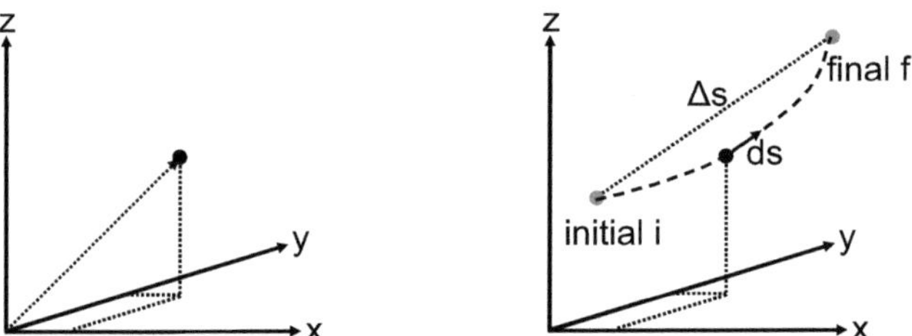

Fig. 1-2 Cartesian coordinate system, left giving a location marked by coordinates (x, y, z) or equivalently a vector from the point of origin, and right a change of position by motion

A **change of position** comprises a **change of location** e.g. from an initial point (x_1, y_1, z_1) to a final point (x_2, y_2, z_2), called **translation**, and a **change of orientation**, called **rotation**, e.g. from an initial orientation $(\alpha_1, \beta_1, \gamma_1)$ to a final $(\alpha_2, \beta_2, \gamma_2)$. For example, somebody moves from one corner of the market place to the opposite corner in a straight line is a linear translation.

Differential changes (infinitesimal differences) are, as is common practice, denoted by "d", while larger differences are denoted by "Δ". For example (Fig. 1-2, right), ds of a path and Δs between its initial and final point.

A change of position is by **motion**, being the general term for movement. Motion can be described in detail e.g. giving the full **path** (Fig. 1-2, right), just by the distance covered between initial and final position, giving local or average velocity etc. The **velocity** v is the change of location with time

$$v = (s_2 - s_1)/(t_2 - t_1) \quad \text{or} \quad v = ds/dt.$$

Eq. 1-1

It can be given as the ratio of finite differences or differential differences. For example, at a velocity of 10m/s a distance of 10m is covered in 1s. The velocity can also be given as a vector $\vec{v}$, comprising the magnitude and its direction, e.g. an apple falls at 10m/s towards the ground. The change of velocity with time is called **acceleration** a

$$a = (v_2 - v_1)/(t_2 - t_1) \quad \text{or} \quad a = dv/dt.$$

Eq. 1-2

For example, at an acceleration of $a = 10 \text{ m/s}^2$ a velocity of 10 m/s is reached starting from 0 m/s after 1s, and 100 m/s after 10 s.

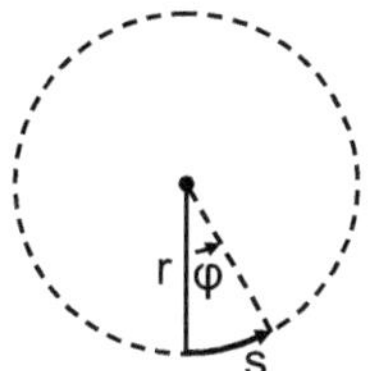

Fig. 1-3 Radius, section of the circumference, and corresponding angle

Changes of orientation are treated similar as of location. If the orientation changes in a plane (Fig. 1-3), a single angle φ is usually used for the description. The change of orientation with time is called **angular velocity** ω

$$\omega = (\varphi_2 - \varphi_1)/(t_2 - t_1) \quad \text{or} \quad \omega = d\varphi/dt.$$

Eq. 1-3

For example, a wheel rotates at 10 rotations, thus 10 times 360°, per second.

From basic geometry (Fig. 1-3) it is known that

$$s = r \cdot \varphi.$$

Eq. 1-4

Then, the velocity is related to the angular velocity by

$$v = ds/dt = r \cdot d\varphi/dt = r \cdot \omega\,.$$

Eq. 1-5

The distance after a full rotation is equal to the circumference $2\pi{\cdot}r$, and thus approximately equal to $6{\cdot}r$. Then, for example, for a radius of 0.5 m, the 10 full rotations per second correspond to a velocity of approximately 10 times $6{\cdot}0.5$ m per second, and thus 30 m/s.

The change of angular velocity with time is called **angular acceleration** α

$$\alpha = (\omega_2 - \omega_1)/(t_2 - t_1) \quad \text{or} \quad \alpha = d\omega/dt\,.$$

Eq. 1-6

Examples of rotation are the motion of a wheel around its own axis, the same for the earth around its north-south axis, but also the movement of the earth around the sun.

We know now how to describe the position and motion of objects in space and time, as well as their changes.

But why do things change? Forces relate to why and how things change, specifically the position of objects (including small particles) by motion (change of position). The forces are due to interactions between objects. Forces and interactions are part of the basics of energy and will be central in the following chapter.

2 Simple situations with 1 or 2 objects

Chapter 2 discusses **simple situations**, which is **1 object in motion or in interaction with a 2nd object**. It starts with a simplified discussion of kinetic and gravitational energy of point-like objects to introduce basic terms, concepts, and laws. In section 2.2, all fundamental energy contributions are discussed, taking also into account effects due to an extension of objects. Simple situations are easy to understand, and their discussion is the basis to understand complex situations, which are then discussed in chapter 3.

2.1 Basic terms, concepts, and laws

Section 2.1 introduces the basic energy related terms, concepts, and laws, specifically force, Newton's laws of motion, momentum and its conservation, energy and its conservation, and energy contributions / forms, conversion between them, their storage, and transport.

For this, simple situations with 1 object in motion or in interaction with a 2nd object, both point-like, having kinetic and gravitational energy, are used. Such situations can be easily observed and are common experience. They are also close to the historic development of the concepts, and allow discussing common misconceptions that occurred in history and still today, and how they are treated and corrected in modern concepts.

We start looking at the term "force", including Newton's laws, because it is a necessary basis to understand "energy", and it is familiar from common experience. The discussion then continues with the term "linear momentum"; it is linked to force in a similar manner as energy, and it is subject to a conservation law, like energy, even having similar origins. The conservation of linear momentum, can (within the validity of these laws) be derived from Newton's 3rd law in connection with his 2nd law, and is first used as an aid to understand the origin and basis of conservation of linear momentum, and for a comprehensive definition of linear momentum. Then, the same is repeated for energy, where the derivation of energy conservation is again (within the validity of these laws) from Newton's 3rd law in connection with his 2nd law. In chapter 3 we will see that this derivation also applies to complex situations. Special to energy, compared to momentum, is that it is has different contributions / forms, e.g. kinetic and gravitational energy, and that conversion between them is possible. The energy contributions / forms, how conversion between them is possible, and why energy is conserved is central in the use of energy and in understanding what energy is.

2.1.1 Forces, motion, and Newton's laws

Newton's laws, called the **laws of motion**, which he published in 1687 in his book "Philosophiæ Naturalis Principia Mathematica" mathematically describe the motion and change of motion of objects having a mass, including the effect of forces on them and between them. They are based on observations of other scientists before him, e.g. Galileo Galilei and Huygens.

A ball is an object of common life, where everybody has experience, and where it is easy to do experiments. We will thus use it as a kind of standard object for the discussion, and for the moment neglect its extension and instead assume it is point-like. What happens in different situations regarding its position and motion? To make things easier to understand, the discussion stays with "common" experience and linear motion as long as possible.

2.1.1.1 Force, inertia, and Newton's 1st law (law of inertia)

The term **"force"** is colloquially used in many ways, e.g. associated with a strength or power of body or mind, physical power, a "push or a pull", and specifically as what causes or can cause changes, like a body to move, or people to do something. Thus, people always used the term "force", and they still do it today, without systematic thought about or analysis of forces. The first evidence of systematic analysis of how forces act on objects comes from ancient Greece: Aristotle investigated forces acting on a balance; later, Galilei made experiments with balls rolling down slopes.

Let's start with the situations shown in Fig. 2-1: a ball on a flat table and the corresponding observations and conclusions.

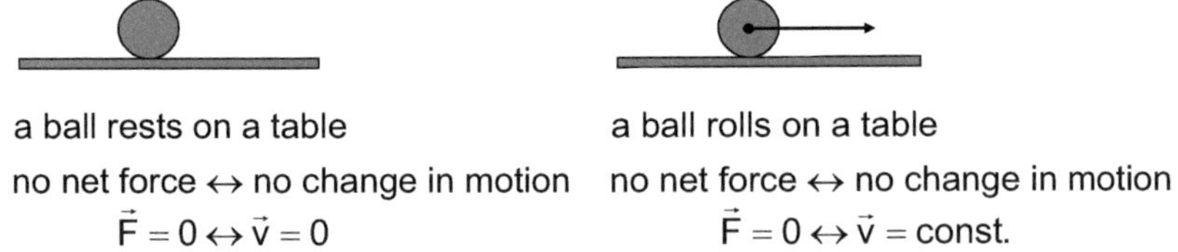

Fig. 2-1 Situations (static and dynamic) with no change in motion

On the left the ball rests. It is common experience that if no force is applied to the ball it stays at rest. The same holds if equal and opposite forces are applied, e.g. if two people try to push the ball from opposite sides; then, no net force is applied as the sum of the applied forces balance, becoming zero.

On the right the ball rolls. It is common experience that a ball rolling on a flat surface will sooner or later come to rest. Similar, when biking, a constant effort is needed to keep the velocity constant, and the same when a horse pulls a carriage. Thus, it is common experience that a force is needed to sustain motion. And this is also what Aristotle believed, and what people believed until Galilei questioned it. He made experiments with balls rolling down even shallower slopes and came to a different conclusion: without a force to stop it, a ball would roll forever; he attributed the difference in observation to the external effect of **friction**. For example, the relation between force and motion in pulling a carriage is different if the ground is ice, solid road, or mud. When friction is reduced, also the necessary force is reduced, and when friction is eliminated the force becomes zero. Friction is not inherent in the motion of the object; its elimination leads to what was already concluded by Galilei, and then expressed by Newton in his 1st law. Objects resist a change in their motion; if there is no (net) force, then there is no change in motion. If no (net) force acts on an object, an object at rest will remain at rest, and an object moving with constant velocity and direction will continue to do so. This is expressed by **Newton's 1st law**, also called **law of inertia**

$$\vec{F} = 0 \leftrightarrow d\vec{v}/dt \equiv \vec{v} = \text{const.} \equiv \vec{a} = 0 \,. \qquad \textbf{Eq. 2-1}$$

An object at rest, where $v = 0$, is a special case of $\vec{v} = $ const. And $\vec{v} = $ const. is equivalent ("$\equiv$") to $d\vec{v}/dt = 0$ or zero acceleration $\vec{a}$. The general meaning of **inertia** is as "resistance to change", e.g. in chemistry inert substances resist a change in their composition. Thus, in common words, a **force** is a "push or a pull", and is the cause of changes in the motion of an object; and no change in motion without a force, thus the "$\leftrightarrow$" sign. If no change in motion is observed, then it can be concluded that there is no (net) force acting. The change in motion is a direct observation, while the force is the concept to explain the observation; both sides refer to different things. So, no net force (explanation) $\leftrightarrow$ no change in the motion (observation). A force is never observed directly; observed is its effect on something.

Newton's 1st law is only valid in what is called an **inertial reference frame**, that is a reference frame that is at rest or moves at constant velocity (including no rotation). In an accelerating reference frame, like an accelerating car, you experience a force pushing you into the seat that disappears as soon as the velocity becomes constant; and the same effect shows up in a curve.

It is important to stress again that the understanding of how forces affect motion has changed throughout history. Common, direct observation is that a force is needed to sustain motion, and this is what ancient philosophers and early scientists believed. The correct answer is found only by making experiments to exclude effects that lead to a wrong outcome. It is important to realize for teaching that the problems people have when starting with the subject today are the same as the problems that were encountered in the historic development. If just the final, correct solution is presented, without elimination of the wrong / old belief by discussion and explanation, then it does not result in understanding the matter. It is merely a new belief replacing the old one, with unanswered questions remaining. It is thus crucial to discuss the historic concepts, their development, and reasons for changes.

2.1.1.2 Free fall and the force of gravitation

Observations of a ball rolling on a flat table are usually made in experiments, but that a ball or another object falls down is something common, and of course was known by people long before philosophers started to think about why objects fall down. The explanation early philosophers gave is gravitation; it acts "downwards", and also explains why people can stand on the ground; already Aristotle discussed it (McPhee 2015). Looking again at the situation where a ball rests on a table, when removing the table, the observation is sketched in Fig. 2-2. The ball falls due to the **force of gravitation** F_g. The "cause" of the force is **gravitation**. Before it must have been balanced by a force due to the table, due to its **interaction** with the ball.

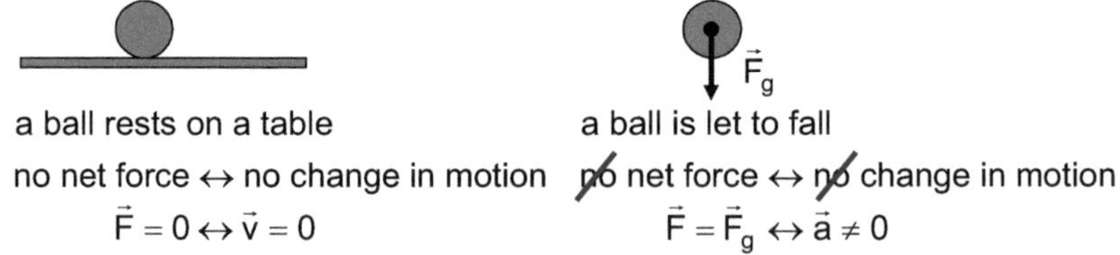

Fig. 2-2 Free fall and the force of gravitation

2.1.1.3 Balance of forces in the presence of gravitation

What if the ball is held motionless by hand? According Newton's 1st law, the force of the hand must balance the force of gravitation such that the net force is zero (Fig. 2-3). It is the same as before for the table.

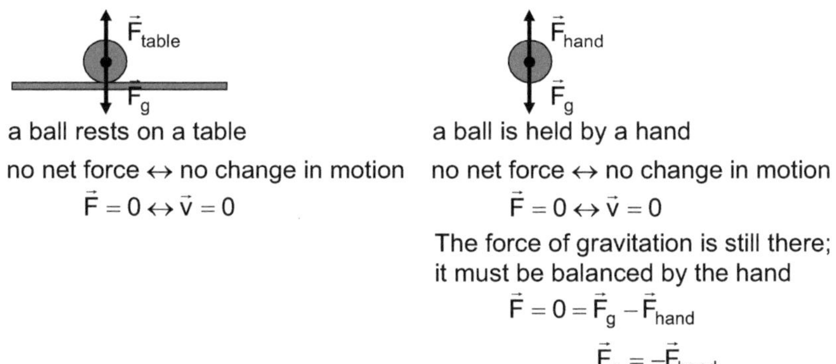

Fig. 2-3 Balance of forces by a table (left) and by hand (right)

In general, for a situation to be motionless, also called static, all forces must balance such that the net force is zero. The field of studying motionless situations is called **statics**, while for situations with motion it is **dynamics**. In architecture, it is important that all forces in a building balance at any point and in all directions (Fig. 2-4); if not, motion starts somewhere in the structure. This was especially critical in ancient times when buildings were made of stones without mortar, where motion can readily lead to a collapse.

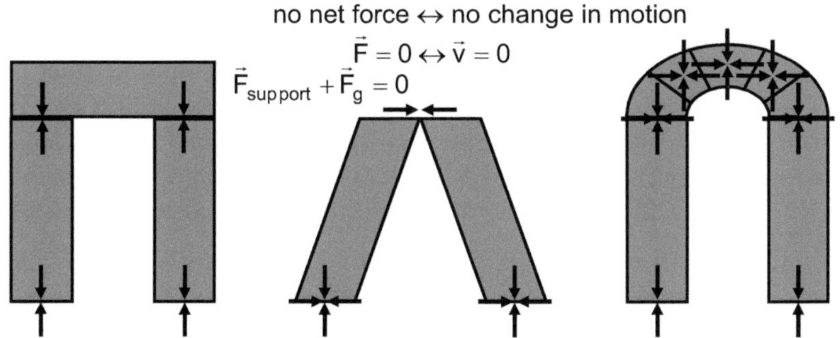

Fig. 2-4 Static situations in architecture, showing the balance of forces

2.1.1.4   Mass and Newton's 2nd law (principle of action)

Newton's 1st law, the law of inertia, states that if no force acts on an object it keeps its motion, but the law does not explain why. What is behind inertia, meaning the resistance of the object to change its motion? And what is the relation between force and change in motion if the force is not zero?

Looking again at the previous situations more closely gives the answer. When a ball is held motionless by hand, F_{hand} and thus also F_g scales with the ball "size", or better what is called its **mass** m, with SI unit $[m] = kg$; heavy objects have a large mass. The force needed to accelerate a ball also scales with the ball "size", or better its mass; heavy objects are hard to accelerate such that the mass also shows up in the description of the dynamic situation. This expresses **Newton's 2nd law**, also called **principle of action**

$$\vec{F} = d(m \cdot \vec{v})/dt ; \quad \text{if } m = \text{const., then } \vec{F} = m \cdot d\vec{v}/dt = m \cdot \vec{a} . \quad \textbf{Eq. 2-2}$$

The force (explanation) and the acceleration (observation) are now connected by the origin of inertia, the objects mass, to a quantitative statement. Newton's 2nd law includes the qualitative statement of Newton's 1st law (law of inertia, Eq. 2-1) saying no net force $\leftrightarrow$ no change in motion. Like Newton's 1st law, the 2nd law holds also only in inertial reference frames. Crucial to understand the nature of forces is that we experience forces only by the effect that they have. A force is never observed itself; "force" is a concept to explain and describe an **interaction** how e.g. a hand, a table, or gravitation interact and affect e.g. a ball. From Eq. 2-2 follows that all forces have the SI unit $[F] = kg \cdot m/s^2 = N$, named after Newton.

This can now be used for a comprehensive **sci.-tech. definition of "force"** as **"A force, in common words a "push or a pull", is a concept introduced in science to explain the cause of changes in the motion of an object (no change without a force). The change of the velocity with time (acceleration) of an object multiplied by its mass is equal to the applied net force (Newton's 2nd law). A force is a vector, thus having a magnitude (quantity) and a direction (its magnitude has the SI unit Newton)."**

Compared to the sci.-tech. definition the term "force" is commonly used in a much broader meaning. The terms force, power (sections 2.1.5), and pressure (section 3.3.2.3), which in science and engineering have own, different meanings, are in common usage often mixed, leading to misconceptions.

2.1.1.5 Gravitational acceleration

An important application of Newton's 2nd law is the free fall of an object due to the force of gravitation. Aristotle concluded from common observation that heavy bodies fall faster than light ones, e.g. a piece of lead falls faster than a piece of wood or a feather. Galilei questioned if this is correct. However, he had no reliable way to measure short times for measurements; one of the methods at hand was to use his own pulse, another the swinging period of a pendulum. Thus, instead of making experiments with free fall, he decided to slow down the fall by letting balls roll down slopes. He observed that light and heavy balls roll the same way and concluded that gravitation works the same on both. That feathers fall slow he explained by friction with air, and thus an external effect. The effect is used in a parachute: despite the additional weight of the parachute the person falls slower.

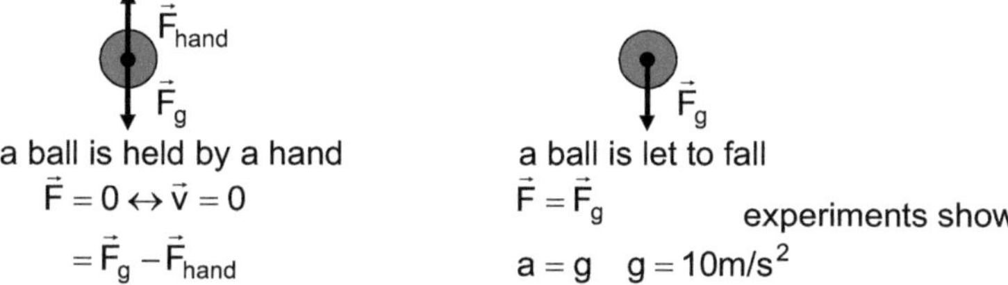

Fig. 2-5 Free fall of an object, e.g. of a ball

Today, experiments on free fall can be made in an evacuated tube such that there is no friction with air, and it can be observed directly that all objects fall the same way, specifically, all fall with the same acceleration (Fig. 2-5), called **gravitational acceleration** g. From Newton's 2nd law follows

$$\vec{F}_g = m \cdot \vec{g}.$$

Eq. 2-3

The value of g is about 10 m/s^2 = 10 N/kg on the earth surface at any point. On the moon the value is however different. We already said that the force of gravitation describes how gravitation affects an object, so obviously there is a difference about the source of gravitation on the earth and on the moon. We will look at that in section 2.2.2 in more detail. For the moment, the important aspect is that the force of gravitation acting on an object is due to an **interaction** with another object, e.g. the earth. This leads to what is probably the single most important issue in this book: Newton's 3rd law.

2.1.1.6 Newton's 3rd law (principle of reaction)

Newton's 2nd law connects the force acting on an object with the change of its motion. But where do forces come from? In general, forces are due to an interaction with a second object. For example, if person A pushes person B, then A also feels a push from B; consequently, A needs to find a position to stand firmly. The same holds if both pull on a rope in opposite directions. Therefore, every interaction between two objects is associated with an action and a reaction force. What is the magnitude and direction of the forces?

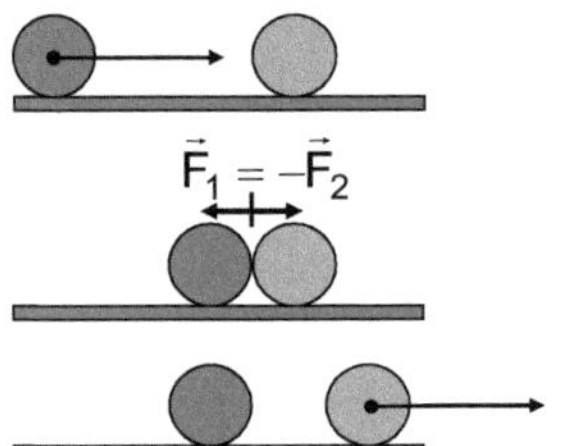

a ball rolls on a table towards another ball of the same mass at rest

after collision, the first ball is at rest while the second ball now rolls with the same velocity that the first one initially had

Fig. 2-6 Collision (elastic) of two balls of equal mass, one ball at rest

Let's look at two balls that interact by an elastic collision (no losses occur, as discussed later in detail); both have equal mass, one is initially at rest. The observation is that the velocity v is conserved (Fig. 2-6). **Conservation means something does not change with time**, e.g. food conservation. The observation can be explained using Newton's 2nd law: on collision, a force $F_1 = m_1 \cdot a_1$ decreases the velocity of the first ball to zero, while a force $F_2 = m_2 \cdot a_2$ increases the second balls velocity to the same velocity as the first ball initially had. Here, because $m_1 = m_2$ and $a_1 = -a_2$, follows $F_1 = -F_2$. And this holds in general, as experiments with different masses m and / or velocities v show. And it is no surprise. When two people pull on a rope in opposite directions the rope becomes like a straight line; thinking of the rope as pulling on the people it becomes clear that both forces have opposite direction and the same magnitude (as long as the rope is not moving itself). This expresses **Newton's 3rd law**, also called **principle of reaction**

$$\vec{F}_{(2 \to)1} = -\vec{F}_{(1 \to)2} \cdot \qquad \textbf{Eq. 2-4}$$

So, a force acting on an object is due to an **interaction** with a second object, which itself is subject to a force of same magnitude but opposite direction. Like rowing a boat, water pushed backward makes the boat move forward.

Newton's three laws, called the **laws of motion** and published in 1687, mathematically describe the motion and change of motion of objects having mass when forces act on them, as well as the forces between them. They allow describing the behavior of single and many objects, large and small …

But Newton's 3rd law is special. Laws stating that something is conserved, e.g. momentum or energy, are quite useful; they will be discussed soon in sections 2.1.2 and 2.1.3. Even Newton's 3rd law (Eq. 2-4) rewritten becomes

$$\vec{F}_1 + \vec{F}_2 = 0, \qquad\qquad \textbf{Eq. 2-5}$$

saying that the sum of the forces in an interaction of two objects is zero. Always. Thus, we could also say that the sum of the forces is "conserved". Newton's 3rd law in its original (Eq. 2-4) or rewritten form (Eq. 2-5) focuses on the interaction between two objects, and is the key to understand energy.

2.1.2 Linear momentum, its conservation, and impulse

We continue the discussion now with the term "linear momentum"; it is linked to force in a similar manner as energy, and it is subject to a conservation law, also like energy, even having similar origins. We will thus use it as an aid to understand the origin and basis of conservation, for a comprehensive definition of linear momentum, and afterwards, do the same for energy.

2.1.2.1 Linear momentum

Newton's 2nd law (Eq. 2-2), as originally published, refers to a term that is the product of an objects mass and velocity, called **linear momentum** p

$$\vec{p} = m \cdot \vec{v}. \qquad\qquad \textbf{Eq. 2-6}$$

It is again a vector, with SI unit $[p] = kg \cdot m/s$. What is its special meaning? While inertia (mass) explains why a body continues to move, e.g. a stone thrown or an arrow shot, momentum explains for how long even if there is an opposing force. Newton's 2nd law states that the change of momentum is due to a force acting on the object for a time. The common use of "**momentum**", e.g. when saying that a group of people gain or loose momentum, or an avalanche coming down a mountain, includes the mass / inertia and the velocity / motion and expresses that both combined result in something that is hard to stop. Common and scientific meaning are thus quite similar.

2.1.2.2 Impulse - linear momentum theorem

Newton's 2nd law (Eq. 2-2) can be rearranged to

$$\vec{F} \cdot dt = d(m \cdot \vec{v}) = d\vec{p} , \qquad \text{Eq. 2-7}$$

and by integration from an initial to a final time gives

$$I = \int_{i}^{f} \vec{F} \cdot dt = \Delta(m \cdot v) = \Delta p , \qquad \text{Eq. 2-8}$$

which is called **impulse - linear momentum theorem**; the left side is called **impulse** I. It states that the change of the linear momentum of an object (observation) is equal to the impulse that acts on the object (explanation, as it includes F). Its importance comes from explaining where the momentum of a flying stone or arrow comes from. It comes from the force that has acted for some time on the stone or arrow before it flies independently.

But for such a force a second object is necessary according Newton's 3rd law. So let's discuss next the situation with two interacting objects.

2.1.2.3 Conservation of linear momentum

If two objects interact, e.g. by a collision (elastic), the forces acting on them have the same magnitude and opposite direction, as Newton's 3rd law states. Using the fact that Newton's 3rd law (Eq. 2-4) is valid at any time gives

$$I_1 = \int_{i}^{f} \vec{F}_1 \cdot dt = -\int_{i}^{f} \vec{F}_2 \cdot dt = -I_2 , \qquad \text{Eq. 2-9}$$

and consequently, using Eq. 2-8, also

$$\Delta \vec{p}_1 = \int_{i}^{f} \vec{F}_1 \cdot dt = -\int_{i}^{f} \vec{F}_2 \cdot dt = -\Delta \vec{p}_2 . \qquad \text{Eq. 2-10}$$

The impulse of both objects has the same magnitude and opposite sign; together both cancel out. And the same holds for the change of the linear momentum. One goes up the same way as the other down. The interaction is thus often described as a "transfer" or "exchange" of linear momentum between both objects (observation) by means of an impulse (explanation).

Using the fact that integration adds an integration constant Eq. 2-10 gives

$$\vec{p}_1 + \vec{p}_2 = \text{const.},$$

Eq. 2-11

or equivalently, using Eq. 2-6

$$m_1 \cdot \vec{v}_1 + m_2 \cdot \vec{v}_2 = \text{const.}$$

Eq. 2-12

Thus, when two objects interact the vector sum of their linear momentum is conserved. This is one of two observations, shown in Fig. 2-7, published already in 1669 by C. Huygens in his "Laws of collision" (Wikipedia 2016b).

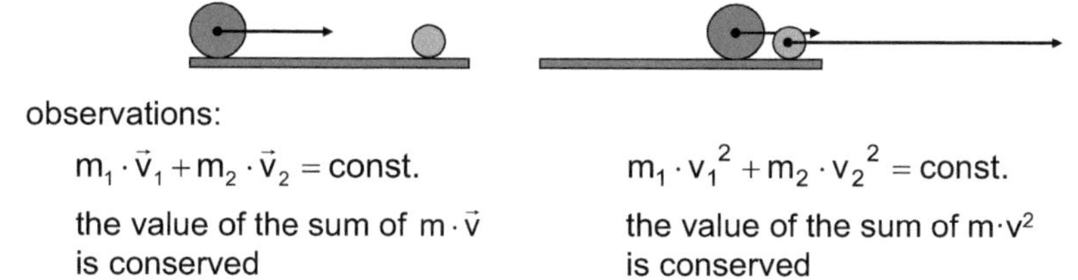

observations:

$$m_1 \cdot \vec{v}_1 + m_2 \cdot \vec{v}_2 = \text{const.}$$

the value of the sum of $m \cdot \vec{v}$
is conserved

$$m_1 \cdot v_1^2 + m_2 \cdot v_2^2 = \text{const.}$$

the value of the sum of $m \cdot v^2$
is conserved

Fig. 2-7 Collision (elastic) of two balls, and observations

The observation is called the **conservation of linear momentum**, and it is obvious that it holds for any number of objects as the collisions can always be seen as pairs of collisions in series or in parallel.

The possibility to derive the conservation of linear momentum from Newton's 2nd and 3rd law of motion is well known (e.g. Kurzweil et al. 2009). The key is that in a single interaction the reaction force and the action force are equal in magnitude and opposite in direction (Newton's 3rd law), that this holds at any time (Eq. 2-10), and that the effect on many objects is already due to the single interaction. This is "hidden" in Eq. 2-11 or Eq. 2-12.

The proof is at this point general within the validity of Newton's laws. There is no reference or limitation on the type of force. A collision of two objects, where actually F is not specified except that it is repulsive and short range, is covered by the proof. The meaning of "**elastic**" is that no losses occur, and is explained in section 3.3. The conservation of linear momentum also holds in cases where Newton's laws are not applicable, e.g. in quantum mechanics or when photons interact. Consequently, the proof presented here is a proof within the validity of Newton's laws and a tool for explaining the general case, but it is not a proof of the general case.

The discussion now allows to make a comprehensive **sci.-tech. definition of "linear momentum"** as **"Linear momentum is a concept introduced in science, used as a measure for the motion of an object, specifically its resistance to change its motion (as described by Newton's 2nd law): the change of the linear momentum of an object, equal to its mass times its velocity, is equal to the applied net force multiplied by the applied time (called impulse). The linear momentum is a vector, thus having a magnitude (quantity) and a direction (its magnitude has the SI unit kg·m/s").** In addition **"The changes of linear momentum of more than one object interacting with each other are interrelated, and cancel out (as can be derived from Newton's 3rd law). Consequently, the vector sum of the linear momentum of all objects is constant, which is called conservation of linear momentum."**

Due to its importance, the conservation of linear momentum should be included in a comprehensive definition because it adds crucial information.

The conservation of linear momentum can be derived from Newton's laws, thus the motion of objects can be described by linear momentum and its conservation as well as by Newton's laws and forces; the result is the same. However, in situations with many objects it is often much more convenient to use the conservation of linear momentum. For example, a box filled with gas; the number of particles in it practically cannot be counted, and the individual application of Newton's laws to all particles and interactions between them is impossible. Nevertheless, when the box is initially at rest, it will stay at rest, no matter if the gas particles collide with each other or the walls of the box. This is obvious without any calculation because the linear momentum in all directions is conserved. The conservation of linear momentum, derived from a simple situation, thus allows predicting the outcome of a complex situation, while a detailed calculation is impossible.

We could continue now discussing **angular momentum** in a similar manner, but it would not add significant information to understand what energy is and why it is conserved. It is later briefly introduced in section 2.2.1.2.

The part "The changes of linear momentum of more than one object interacting with each other are interrelated, and cancel out ..." resembles what comes next: the conservation of energy. Energy conservation is central in the use of energy, and in understanding what energy is. This is the reason why we have discussed linear momentum here before energy.

2.1.3 Energy, its conservation, and work

The collision of a ball with another ball, shown in Fig. 2-7, already reveals that not only the linear momentum is conserved; there is another value, $m \cdot v^2$, which is also conserved. This observation, not discussed or explained yet, is part of our next focus. It will lead to something that is even more useful than the linear momentum and its conservation: energy.

2.1.3.1 Kinetic energy and gravitational energy

The observation of the collision of two balls in Fig. 2-7 is for a flat table, meaning for constant height. What happens if the height h is not constant? This can be investigated by looking at the free fall of a ball, or when a ball rolls down a slope, but the case when a ball is part of a pendulum readily gives more insight. Fig. 2-8 shows this and the crucial observations.

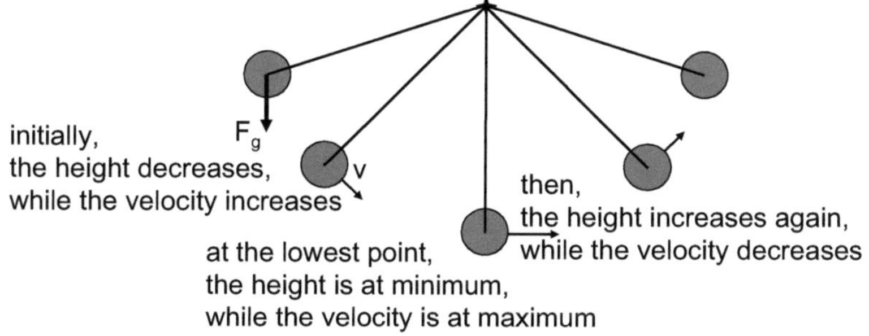

Fig. 2-8 Ball as part of a pendulum, and the crucial observations

When the pendulum swings, it swings forward and backward, again, and again. The repetition of the motion shows us that something is "conserved". Strictly speaking, like with the rolling ball on a flat surface, friction will eventually stop the motion, but as we are already aware that it is an external effect, we neglect it now. We also observe that when the height of the ball decreases its velocity increases and then reverse; in simple words: height (location) is "converted" to velocity (motion) and then back again. Already Galilei observed "some kind" of conversion when a pendulum swings, and this is the key to understand the importance of the term $m \cdot v^2$.

So, what is it specifically that is conserved, and what exactly is converted?

The answer is the sum of m·v^2 multiplied by ½ and m·g·h is constant

$$\frac{1}{2} \cdot m \cdot v^2 + m \cdot g \cdot h = \text{const.}.$$ **Eq. 2-13**

The sum is conserved, and "**conversion**" refers to the observation that one term increases while the other decreases when the pendulum swings. This is described by the term **energy**, which is used in many ways. First, the different contributions are "energy", specifically called **energy contributions**. The sum of all contributions is called **total energy**. That its value is constant is called **conservation of energy**, like for linear momentum. Energy contributions related to motion like ½·m·v^2 are called **kinetic energy** E_{kin}. Those related to position are called **potential energy** E_{pot}. In general, the term "**potential**" means "that can or may come into existence or action, latent capacity" (Hornby 1983); it is a synonym for "possible" or "latent". Thus, while kinetic energy is directly visible due to the associated motion, the term potential energy indicates some "hidden source" of kinetic energy. The term m·g·h is **gravitational energy** E_g, an example of potential energy. All energies have the Si unit [E] = kg·m/s^2 = J, named after J.P. Joule.

Now, how is Eq. 2-13 derived? It can be derived by evaluating experiments. The first step is to plot the observed values of velocity v and height h in a diagram, as in Fig. 2-9; the observation is a square relation: v^2 relates to h.

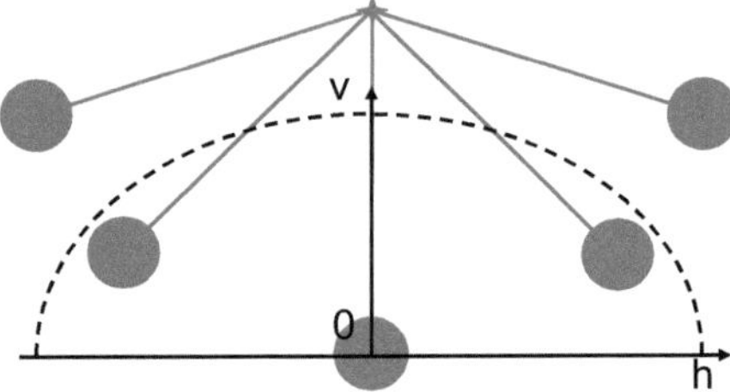

Fig. 2-9 Quantitative analysis of the observations on the pendulum

Thus, the term m·v^2 and not m·v from the h = const. experiments must be combined with m·h, giving ?·m·v^2 + ?·m·h = const. with unknown factors. The experimental data show that the factor is 2·g. This leaves the option where to put it. Multiplication of the second term by g and dividing the first term by 2, giving Eq. 2-13, has the advantage that the gravitational energy and all other potential energies have the form of a force times a distance.

When a ball collides with another, second ball at constant height h it is observed (Fig. 2-7, right) that the sum of "m·v^2" of the ball and the same of the other ball is constant; using the factor ½ again, this can been written as

$$\frac{1}{2} \cdot m_1 \cdot v_1{}^2 + \frac{1}{2} \cdot m_2 \cdot v_2{}^2 = \text{const..}$$

Eq. 2-14

While the conservation of momentum (Eq. 2-12) includes only terms of the form m·v, energy combines terms of ½·m·v^2 (Eq. 2-14) but also terms of m·g·h (Eq. 2-13), thus not only terms relating to motion but also to position. Combining so different things makes energy so hard to understand, but also so useful. The respective equation, called **energy balance**, looks different depending on the situation, e.g. Eq. 2-13 for the pendulum, or Eq. 2-14 for the collision of two balls. And **energy conversion** is between the different contributions. Strictly speaking, there is no "thing" that is "converted"; it is just that one value goes up while the other goes down. Nobody would say that momentum is converted in a collision between two colliding objects.

Fig. 2-10 shows a pendulum interacting with another one by collision, combining the previous examples; left: pendulum 1 starts to swing; center: collision; right: pendulum 2 at maximum height. The energy balance has two terms for kinetic as well as for gravitational energy, one for each pendulum.

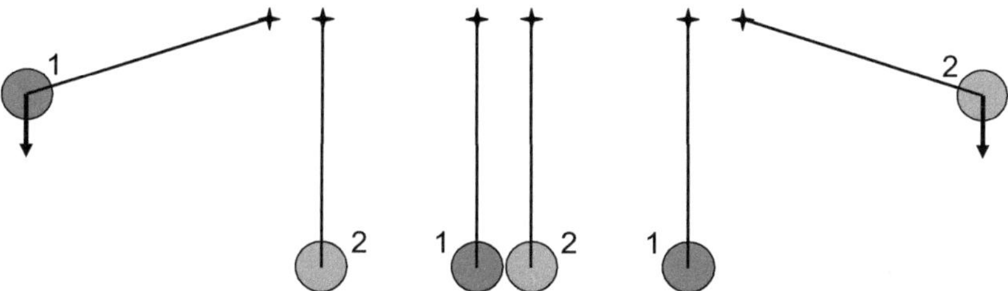

Fig. 2-10 A pendulum interacting with a second one by collision

2.1.3.2 Conservation of energy

To understand the concept of energy better it is necessary to understand energy conservation better, especially its origin and basis. For simple situations, it is in many ways similar to linear momentum and its conservation. Again, the key is that the reaction force and the action force are equal in magnitude and opposite in direction. However, instead of focusing on the time that a force acts in an interaction as with momentum (Eq. 2-9), for energy it is the distance that a force acts. If two objects interact, for example by a collision (elastic), the forces acting on them have the same magnitude and opposite direction, as Newton's 3rd law (Eq. 2-4) states. Using the fact that Newton's 3rd law is valid at any point on a path of interaction means

$$\int_i^f \vec{F}_1 \cdot d\vec{s} = -\int_i^f \vec{F}_2 \cdot d\vec{s}, \qquad \text{Eq. 2-15}$$

or by reorganizing

$$\int_i^f \vec{F}_1 \cdot d\vec{s} + \int_i^f \vec{F}_2 \cdot d\vec{s} = 0. \qquad \text{Eq. 2-16}$$

Thus, the integral of the acting force and the integral of the reacting force along the path of interaction added up gives zero. This is the basis for energy conservation; anything else is related to specific cases and viewpoints.

If the force accelerates the object, then Newton's 2nd law (Eq. 2-2) has to be inserted in one term. Using a = dv/dt and ds = v·dt the term takes the form

$$\vec{F} \cdot d\vec{s} = (m \cdot \vec{a}) \cdot d\vec{s} = m \cdot \frac{d\vec{v}}{dt} \cdot \vec{v} \cdot dt = m \cdot \vec{v} \cdot d\vec{v}$$

$$= d\left(\frac{1}{2} \cdot m \cdot v^2\right) = dE_{kin}. \qquad \text{Eq. 2-17}$$

For example, Eq. 2-16 becomes Eq. 2-13, describing the pendulum, when Eq. 2-17 replaces the first term and in the second term F is the force of gravitation (Eq. 2-3) resulting in m·g·h. And Eq. 2-14, describing when two objects collide, results when applying Eq. 2-17 to both terms in Eq. 2-16.

The proof is at this point general within the validity of Newton's laws. There is no reference or limitation on the type of force or the range, so collision (elastic) of two objects is covered by the proof. The conservation of energy also holds in cases where Newton's laws are not applicable, e.g. in quantum mechanics or when photons interact. Consequently, the proof presented here is a proof within the validity of Newton's laws and a tool for explaining the general case, but it is not a proof of the general case.

Examples

The following examples highlight again a few things that we have just discussed, and point to new things for the further discussion.

First, it is crucial to stress again that in contrast to momentum and its conservation, energy and its conservation combines terms with different form, depending on the situation e.g. $\frac{1}{2}\cdot m\cdot v^2$ and $m\cdot g\cdot h$ instead of only $m\cdot v$. Thus, besides motion also position is involved. It is this difference that makes energy so hard to understand, but also so useful. The respective equation, called energy balance, looks different depending on the situation.

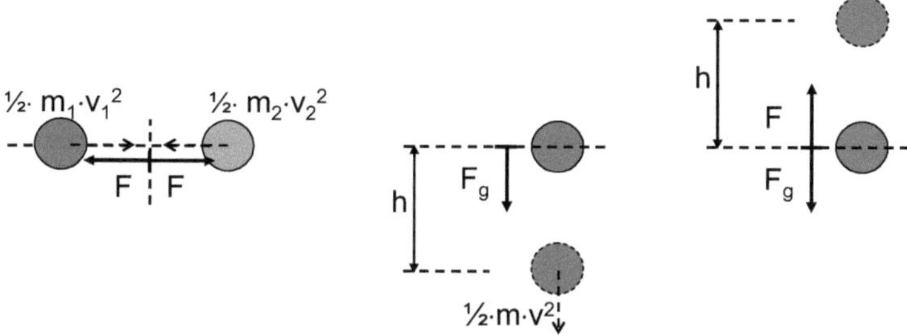

Fig. 2-11 Examples: two objects colliding with each other, an object accelerated due to the force of gravitation and falling a distance h, and a force lifting an object a distance h against the force of gravitation

Fig. 2-11 shows examples of the energy of objects in simple situations, indicating the high variety of energy terms compared to linear momentum. The examples can be described by the equations just derived. Fig. 2-11, left, shows two objects colliding; here, from energy conservation follows that $\frac{1}{2}\cdot m_1\cdot v_1^2 + \frac{1}{2}\cdot m_2\cdot v_2^2 = $ const. applies. Fig. 2-11, centre, shows an object that is subject to the force of gravitation (by interaction with an object not shown) and let to fall free; then, $F_g\cdot h + \frac{1}{2}\cdot m\cdot v^2 = $ const. applies and the kinetic energy increases at the expense of gravitational energy. This is also the case in a pendulum; the rope just forces the object on a circular path with repetition. Finally Fig. 2-11, right, shows a force, e.g. by hand, lifting an object against the force of gravitation; here, $F_g\cdot h + F\cdot h = $ const. applies. The examples show the conversion between kinetic and potential energy, but also that in simple situations it is possible to describe things by forces. It is not surprising now that with such a variety of contributions that are connected to different changes of motion or position, energy is very useful.

2.1.3.3 Historic roots of "energy" – the "golden rule of mechanics"

Before continuing, it is helpful to look at the historic roots of the concept of "energy" and see how it has evolved through history for simple situations.

People tried early to make life easier. Many things were known from experience by try and error, long before an "explanation" was developed. It is not surprising that the force and its optimization was the first step. How is it possible to lift a heavy stone? This question arises long before thinking about how to move the stone more efficiently. Already in ancient times, people used levers and other tools without being aware of the laws behind.

The first investigations on forces are attributed to Aristotle (384 to 322 BC). By observation of a balance (Fig. 2-12, left) with lever arms of different length l, he realized that levers allow moving great weights with little force.

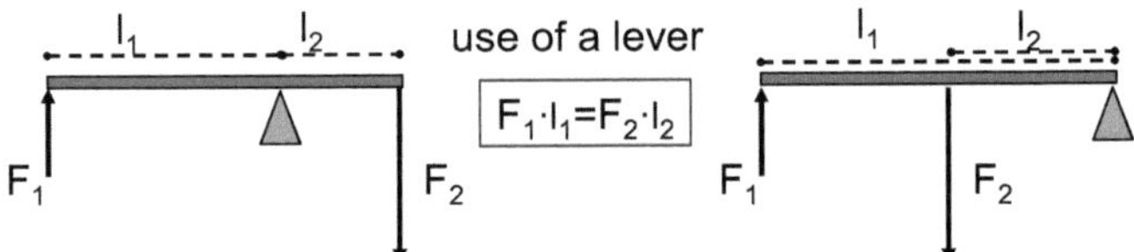

Fig. 2-12 Use of a lever, to "transform" small to large forces

Later, Archimedes (287 to 212 BC) stated and proved that weight times lever arm length is the same on both sides, giving a quantitative relation, not just a general one. The connection of force and length of the lever arms is

$$F_1 \cdot l_1 = F_2 \cdot l_2 \quad \text{or} \quad F_1 / F_2 = l_2 / l_1. \qquad \text{Eq. 2-18}$$

Archimedes described the lever as tool to "transform" small to large forces, using lever arms of different length. He supposedly said that if he had a good enough lever and a place to stand, he could move the earth. Besides the lever, block and tackle (Fig. 2-13) was invented for the same purpose.

But people continued to optimize things, looking not only at the force but also the force times the distance it acts (Fig. 2-13). Corresponding mechanisms allow to "transform" the force times the distance it acts, and adapt it to different needs. On the driving side of the mechanism acts the driving force for the driving distance, while on the load side the transformed load force acts for the load distance, for example to lift a heavy weight.

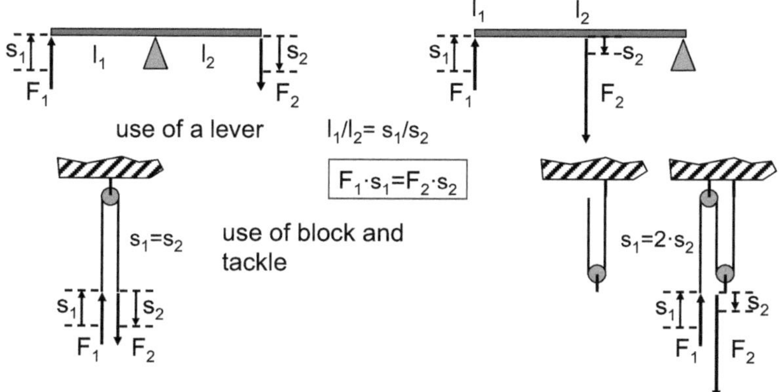

Fig. 2-13 Use of the lever, or block and tackle, to "transform" a force times distance it acts

The value of force F times the distance s it acts is the same anytime

$$F_1 \cdot s_1 = F_2 \cdot s_2 .$$
Eq. 2-19

This expresses the so-called **"golden rule of mechanics"**, formulated by Galilei 1594; it states that **"the driving force times the driving distance is equal to the load times the load distance"** or, in a simple non-quantitative way that "if the force is decreased, the distance is increased".

The golden rule expresses energy conservation; taking into account that in Eq. 2-19 only the magnitude of the forces is considered, which actually have opposite sign, Eq. 2-19 is equivalent to Eq. 2-16. But as said before, for these simple situations nobody thought about "energy" and its "conservation"; using the concept of force is much more straightforward.

Crucial is however the following. Archimedes just focused on the force, with the lever arms being variables. Galilei with the golden rule then looked at the force and distance it acts. Both put their observations into formulas called laws or rules, but there was so far no understanding of the basics. Newton chose a significantly different point of view: he looked at interactions at a single point / object. This corresponds to eliminating the lever arm (or any equivalent) and its effect from the problem. Then, the distances are equal, and the forces become equal in magnitude and opposite in direction. Thus, the basis of energy conservation becomes visible: Newton's 3rd law.

2.1.3.4 Work - kinetic energy theorem

As Eq. 2-17 resembles Eq. 2-7, there is an equivalent to the impulse - linear momentum theorem (Eq. 2-8) for energy. By integrating Eq. 2-17 follows

$$W = \int_{i}^{f} \vec{F} \cdot d\vec{s} = \Delta\left(\frac{1}{2} \cdot m \cdot v^2\right) = \Delta E_{kin}, \qquad \text{Eq. 2-20}$$

called **work - kinetic energy theorem**; the left side is the **work** W. Thus, the change of the kinetic energy of an object is equal to the work done on or by the object. As a consequence of Newton's 2nd law it holds only in inertial reference frames. Like the impulse - linear momentum theorem, it connects an explanation, the work, with an observation, the change in kinetic energy.

The collision of two objects can now be described using Eq. 2-20 as

$$\Delta E_{kin,1} = W_1 = \int_{i}^{f} \vec{F}_1 \cdot d\vec{s} = -\int_{i}^{f} \vec{F}_2 \cdot d\vec{s} = -W_2 = -\Delta E_{kin,2}. \qquad \text{Eq. 2-21}$$

The work on both objects has the same magnitude and opposite sign, thus together both cancel out. The same holds for the change of kinetic energy. One goes up the same way as the other down. The process is often called a "transfer" or "exchange" of energy between both objects (observation) by means of work (explanation). Work stands for the energy "exchanged".

2.1.3.5 Work - energy theorem

If an object is lifted, e.g. by hand, the force of the hand balances the force of gravitation, and the acting distance is the change in height; consequently

$$W = \int_{i}^{f} \vec{F} \cdot d\vec{s} = \Delta E_{pot}. \qquad \text{Eq. 2-22}$$

Combining Eq. 2-20 and Eq. 2-22, we can now generalize the work - kinetic energy theorem to the **work - energy theorem** (Halliday et al. 1993)

$$W = \int_{i}^{f} \vec{F} \cdot d\vec{s} = \Delta E = \Delta E_{kin} + \Delta E_{pot}. \qquad \text{Eq. 2-23}$$

The change of the energy of an object is thus equal to the work done on or by the object. Consequently also, Eq. 2-21 extended by potential energy is

$$\Delta E_{kin,1} + \Delta E_{pot,1} = W_1 = \int_i^f \vec{F}_1 \cdot d\vec{s}$$

Eq. 2-24

$$= -\int_i^f \vec{F}_2 \cdot d\vec{s} = -W_2 = -\Delta E_{kin,2} - \Delta E_{pot,2} \ .$$

Thus, work can be done by a change of energy of (1), and in turn, the work done can change the energy of (2). Overall, the changes of all energies in Eq. 2-24 cancel out, like for momentum in Eq. 2-10, and similar to Eq. 2-11 for momentum now energy is conserved

$$E_{kin,1} + E_{pot,1} + E_{kin,2} + E_{pot,2} = const. \ .$$

Eq. 2-25

Because "work" is crucial in the concept of energy, especially its application, let's have a detailed look, starting with technical matters.

First, the work - energy theorem explains why the choice was to put ½ as a factor in kinetic energy; otherwise, it would show up in the work and potential energies, and especially in work as "force times distance" be confusing.

Second, the notation looks a bit strange like with momentum and impulse: on the work side is an absolute value, while on the energy side a difference. The non-differential form follows the wording in the definition in Eq. 2-23 that the change of the energy of an object is equal to the work done on it. The common usage in literature (Fermi 1956, Reif 1985, Atkins 1990, Halliday et al. 1993, Çengel and Boles 2002) is when using differentials to write $dW = F \cdot ds$ or $p \cdot dV$ etc., $W = \int dW$, and $dE = dQ \pm dW$ (heat Q is introduced later, and thermodynamics redefines the sign of work later to "-"), and in non-differential form to write $\Delta E = Q \pm W$.

Moreover, the work - energy theorem refers to the change of the energy, and not its absolute value. Because we are often only interested in the work, which is useful for an application, this is then sufficient. And it is much easier to calculate or measure just differences of energy. However, this does not allow drawing the conclusion that an absolute energy does never exist. For example, the kinetic energy of an object is zero if its velocity is zero.

What is the meaning of work in a sci.-tech. context, and its common usage? Colloquially, "work" is the "use of bodily or mental powers with the purpose of doing or making something; something to be done; product of the intellect or imagination" (Hornby 1983), thus, the action, the job, as well as the result. In a sci.-tech. context, "work" was introduced as "weight lifted through a height" in 1826 by Gaspard-Gustave Coriolis (wikipedia 2017a), based on using steam engines to lift buckets of water out of flooded mines. Not specifying the force is the definition of "work" as "force times displacement" by Sears and Zemansky 2016, like the work - energy theorem. A definition by Atkins 1990 is even more general: "Work is done during a process if that process could be used to bring about a change in height of a weight somewhere in the surroundings", thus including any possible option. Similar, Atkins 1990 states that "Energy is the capacity of a system to do work." This is the original historic meaning, from the Greek term "energeia", composed of "en", meaning "in, within" and "ergon", meaning "work" (English Oxford living dictionaries 2017). Here, energy is only seen as source, but misses energy (changes) as result, e.g. the potential energy of the stone once it is on top of the wall. The definition by Brown et al. 2015 "Work is the energy used to cause an object to move against a force" states work itself is energy. But this is a bit misleading; work is not just energy, like impulse is not momentum. Similarly, Cleveland and Morris 2009 state that „work" is "the useful transfer of energy to a body by the application of a force, causing the body to move a certain distance in the same direction as the applied force (product of force and displacement)"; it again just stress that work ends as energy, but adds "useful". But, if a result is useful or not can be a personal opinion; we often say that "all the work was useless".

Taking all the relevant aspects together now allows making a preliminary comprehensive **sci.-tech. definition of "work"** as follows: **"Work denotes energy exchanged / transferred between an object and its surrounding, specifically by the process of a force that is acting for a distance. Thus, work done on or by an object leads to a change of its energy, thereby lowering or rising its kinetic or potential energy. If the origin and destination of the work are discussed together, then the energy of the origin is lowered while that of the destination increases or reverse."**

Observable are the changes of energy, by a change of velocity or height, while work is the explanation. Later, in chapter 3, we will replace the term "object" by the term "system", and check the preliminary definition again.

With the work - energy theorem the **concept of energy** now starts to enfold. To better understand it, and as an exercise, let's discuss some examples.

For example, a heavy stone on one side of a lever goes down, and thereby lifts another, lighter stone on the other side. Then, the first stone does work to raise the energy of the other stone. The work is the force of gravitation of the first stone times the distance it moves, while the energy change is the force of gravitation of the second stone times the distance it moves; in total $W = m_1 \cdot g \cdot \Delta h_1 = \Delta E = m_2 \cdot g \cdot \Delta h_2$. But we can see it also the opposite way: the energy of the first stone changes while it does work on the second stone: $\Delta E = m_1 \cdot g \cdot \Delta h_1 = W = m_2 \cdot g \cdot \Delta h_2$. Therefore, it depends on the point of view.

Let's now discuss the examples from Fig. 2-11 again as an exercise.

Fig. 2-11, left, shows two objects colliding. Looking at both objects together, from energy conservation follows that $\frac{1}{2} \cdot m_1 \cdot v_1^2 + \frac{1}{2} \cdot m_2 \cdot v_2^2 = \text{const}$. When looking at both objects individually, the explanation by using the work - energy theorem is that the collision is an interaction causing a transfer or exchange of kinetic energy between both objects (observation) by means of work (explanation).

Fig. 2-11, centre, shows an object that is subject to the force of gravitation and let to fall free; the kinetic energy increases at the expense of gravitational energy, and from energy conservation follows $F_g \cdot h + \frac{1}{2} \cdot m \cdot v^2 = \text{const}$. An alternative view from the work - energy theorem is that the force of gravitation does work on the object and thereby increases its kinetic energy.

Fig. 2-11, right, shows a force, e.g. by hand, lifting an object against the force of gravitation. From energy conservation follows $F_g \cdot h + F \cdot h = \text{const}$. An alternative view is that the force does work on the object, thus increasing its gravitational energy.

The examples show several things. First, the force in work can be arbitrary, but also specifically the force of gravitation or any other. Second, force times distance is not always work; it depends on the point of view: the force of gravitation acting along a distance can be work, but when an object is lifted its value also describes its change of gravitational energy. Third, the work - energy theorem (Eq. 2-23) is very useful. But everybody knows that the work done when pulling a stone up a slope depends on friction; the resulting change in gravitational energy is however the same. This will lead us to the final expansion by adding heat and thermal energy in section 3.3.3.

2.1.4 Energy conversion, energy contributions, and energy forms

The examples discussed so far comprise two energy contributions to the total energy; they are kinetic energy due to the motion of an object

$$\frac{1}{2} \cdot m \cdot v^2 = E_{kin} \, ,$$

Eq. 2-26

and gravitational energy, which is a special example of potential energy

$$m \cdot g \cdot h = E_g \, ;$$

Eq. 2-27

it is due to the interaction between an object and the earth, or another mass. Later we will see that there are more. For just kinetic and gravitational energy, e.g. for the pendulum or for the free fall, the total energy is therefore

$$\frac{1}{2} \cdot m \cdot v^2 + m \cdot g \cdot h =$$
$$E_{kin} \quad + \quad E_g \quad = E_{tot} \, .$$

Eq. 2-28

In the examples of free fall or the swing of a pendulum, we already saw that energy in one energy contribution can increase, while decrease in another. The energy contributions are connected, and thereby connect also changes of motion and position, characterized by velocity and height of an object. Energy conservation is connecting these changes in a quantitative relation. This allows making predictions, use them for optimization, and it allows understanding things better. It is why the concept of energy is so useful.

2.1.4.1 Energy contributions and energy forms

Up to now, we have used the term energy contribution and not energy form. Both have a similar meaning, but knowing their difference is quite helpful. For none a definition was found in literature, so they are defined now here. An **energy form** refers to changes that are directly observable, macroscopic, where energy was discovered in history. An **energy contribution** is anything that contributes to energy. For example, kinetic energy is an energy form when it refers to the energy of a moving car, but the kinetic energy of atoms in a stone that we can't see directly is an energy contribution; it contributes to the energy form of thermal energy related to temperature.

2.1.4.2 What is energy conversion?

Now, what is energy conversion? When a pendulum swings, we usually say that gravitational energy is "converted" to kinetic energy, and back again. Actually, the original statement of energy conservation, which is **"energy cannot be created or destroyed; it can only be transformed from one form to another"**, uses the term "transformation" combined with that of energy form. In common literature both, conversion and transformation, as well as energy contribution and energy form are often used synonymously. **Conversion** in general means that something changes to something else; a common example of "conversion" is the conversion in chemical reactions. This is what people had in mind when the term was introduced to the field of energy. But saying that energy is "converted" between different energy contributions / forms can be misleading. Instead of implying that energy is some "thing" like a substance, changing in a process of energy conversion, it is better to say that **energy "conversion" means the amount of energy in one energy contribution / form increases, while the amount of energy in another energy contribution / form decreases.** There is nothing mystical about the conversion / transformation of a compound to another in a chemical reaction; it is just a reorganization of how atoms are bound in the compounds. For energy it is similar. The meaning of energy conversion between different contributions / forms can be understood from the discussed examples of simple situations, where the focus is on a single object and e.g. its interaction with a second object, especially because we can use forces; the use of "energy" there is not necessary. Due to interactions between objects, the energy in the different energy contributions / forms can change. But energy as a concept was not used as long as people looked at simple situations like a pendulum, or a planets motion around the sun; they can be described without using "energy". The term "energy" became common when it became useful, and that was when people could not explain the observations otherwise, e.g. the conversion of "work" to "heat". And here, because energy was "converted" as well as conserved, people started imagining energy as some "thing" like a substance, where mass is conserved and properties change in a chemical reaction. In chapter 3 we will discuss this in more depth, and we will see that we can understand also what is going on in complex situations, based on understanding what is going on in simple situations.

2.1.4.3 Mechanisms and cause of energy conversion

Discussing simple situations allows understanding the mechanisms behind energy conversion, meaning how it works, as well as the cause of energy conversion, meaning why it is happens. So let's have a look at it in detail. For this, we can use the two examples of simple situations discussed up to now where conversion is between gravitational energy and kinetic energy: free fall, and the swing of a pendulum where the conversion is forward and backward repeatedly.

Mechanisms behind energy conversion, meaning how it works

What is the mechanism behind conversion in both cases? It is obviously that $E_g = m \cdot g \cdot h$ and $E_{kin} = \frac{1}{2} \cdot m \cdot v^2$ refer to the same mass, and if the location changes, with v having a component in the direction of h, then both contributions change; they are "connected".

Energy conversion needs a connection between the different energy contributions / forms, e.g. the mass and location of an object connects its kinetic energy and its gravitational energy.

Cause of energy conversion, meaning why it happens

And what causes energy conversion? In both cases, the free fall as well as the swing of the pendulum, conversion starts by the force of gravitation, more general a net force. Thus, it is a net force that leads to the change of location of the object, here F_g, causing acceleration or deceleration and thus an increase or decrease of E_{kin}, while the resulting change in location causes the opposite for E_g. But there is something else about the pendulum; compared to the free fall, why does the pendulum swing, meaning convert E_g to E_{kin} not only one way, but back repeatedly? At the lowest point, h_{min}, nothing would happen if the velocity would be zero; there is no net force. This has actually been a problem of people in the Middle Ages; inertia was already known, but just inertia could not explain why the pendulum continues to swing, because it also has inertia when at rest. What makes the pendulum keep going, and with it energy conversion, is the momentum of the object, thus the product of inertia and velocity. At the lowest point, where there is no force in the direction of motion, the momentum is highest.

Energy conversion to start or proceed needs a net force.

Forces with no effect on energy

Let's now look at the forces when a pendulum swings in more detail. We need to remember that Newton's 2nd law (Eq. 2-2) is a vector equation; force and motion have a direction that must be considered. This is obvious: to lift an object you pull upwards, not sidewards, and to increase the velocity of an object in a direction you pull in that direction, not from the side.

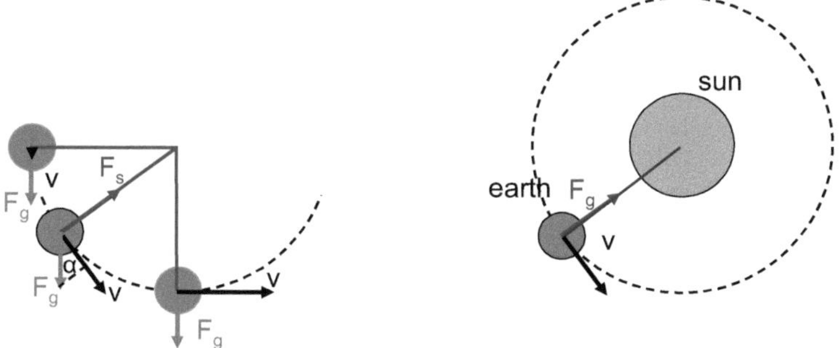

Fig. 2-14 Examples for forces perpendicular to the direction of motion

As Fig. 2-14, left, shows, F_g is initially in direction of motion of the object, thus leading to an acceleration in that direction. At the lowest point F_g still acts downwards while motion is sidewards; thus there is no acceleration due to F_g. And at points between? In general, the part of a force, here F_g, that acts in the direction of motion is its projection on the direction of motion: the product of the magnitude of the force and the cosine of the angle α between force and direction of motion

$$\left|\vec{F}\right| \cdot \cos\alpha \,. \qquad\qquad \text{Eq. 2-29}$$

And energy? Pulling sidewards does not lift an object; it therefore does no work, and therefore also not change its gravitational energy. In general

$$\vec{F} \cdot d\vec{s} = \left|\vec{F}\right| \cdot \left|d\vec{s}\right| \cdot \cos\alpha \,. \qquad\qquad \text{Eq. 2-30}$$

Thus, the kinetic energy in forward direction also does not change when pulling sidewards. E.g. the force of the rope F_s in a pendulum has no effect on energy, and the same for F_g in the earth orbit around the sun (Fig. 2-14).

2.1.5 Energy conversion, storage, and transport

The main points discussed up to now are that energy is overall conserved, that energy appears in different energy contributions / forms while conversion between them is possible, and that energy exchanged between things or transferred from / to things is called work when a force acts a distance. Since "transfer" could imply that the process is done actively, maybe with an intention, the term "exchange" is used throughout this book.

Besides energy conversion there are two more types of processes that are central to the use of energy, especially in energy technology. They are introduced here briefly for completeness. Energy conversion, storage, and transport are together the three basic types of processes that are the basis for whatever happens in the energy system. They are complementary processes (Mehling 2016), and in simple words, energy conversion is affecting the energy contribution / form, energy storage is affecting the time, and energy transport is affecting the location.

Fig. 2-15 shows examples from the simple situations previously discussed.

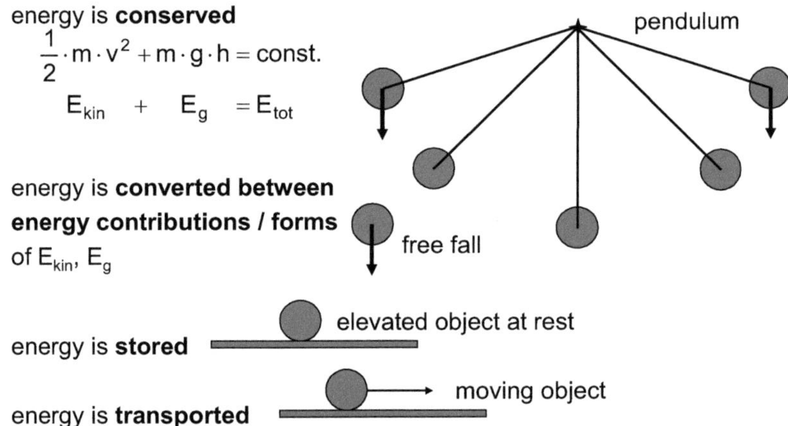

Fig. 2-15 Energy conversion, storage, and transport – examples from simple situations

Let's now discuss and define the three basic types of processes briefly, before continuing.

In common words, conversion means something is changed to something else. A **sci.-tech. definition of energy conversion** is thus **"Energy conversion means the amount of energy in one energy contribution / form increases, while the amount of energy in another decreases"**. The use of the terms energy contribution and energy form, as well as energy conversion and energy transformation, which are often used synonymously, was already discussed in section 2.1.4.1 and 2.1.4.2.

What about storage and transport, not discussed yet?

In common words, storage means keeping something for some time, to use it later. A **sci.-tech. definition of energy storage** is thus (Mehling 2016) **"Energy storage means keeping energy for some time for using it later"**. Strictly speaking, it usually refers to an energy contribution / form.

And in common words, transport means moving something in space, from one location to another location. A **sci.-tech. definition of energy transport** is thus (Mehling 2016) **"Energy transport means moving energy in space, from one location to another location"**. Strictly speaking, it usually also refers to an energy contribution / form. Besides "transport" the term "transfer" is also in use. The term "transport" refers to another location (across a distance), while "transferred" is within a material, or just across a boundary (no distance), or between things. For example, money is transported in a van to another bank, but it is transferred between accounts, which can even be within the same bank (then it is not transported).

Why are energy storage and energy transport not discussed in any detail? The answer is that energy storage or energy transport reveal nothing about energy and its contributions / forms that is not already revealed by looking at energy conversion. Energy storage just preserves a situation of an energy contribution / form, and energy transport stays with the same energy contribution / form, sometimes just moves stored energy.

Finally, one more term used with these processes is missing: power. A quantity of energy per time, for example converted, stored, transported, used etc. is called **power**

$$P = dE / dt \,.$$ **Eq. 2-31**

Its SI unit is $[P] = J \cdot s^{-1} = W$, named by **Watt**. Power and force have different sci.-tech. meaning, but are in common usage often used synonymously.

2.1.6 What is energy, and what is it not?

Section 2.1, discussing simple situations, already revealed a lot on energy, its conversion and conservation, and the relation to interactions, thereby on what energy is and what it is not. Let's summarize the key findings.

What is energy?

A preliminary **comprehensive, sci.-tech. definition of "energy" could be "Energy is a concept introduced in science, to describe the effect of interactions between objects on the motion and position of the objects. This is done by energy contributions / forms, e.g. $\frac{1}{2} \cdot m \cdot v^2$ for the motion of an object, called kinetic energy, or $m \cdot g \cdot h$ for the position of an object that interacts by gravitation with the earth, called a potential energy. Energy is a scalar, thus has only a magnitude (quantity); its SI unit is $kg \cdot m/s^2$."** Crucial is that **"The sum of the energy in all energy contributions / forms of a single object or of more than one object interacting with each other is constant, which is called conservation of energy."** When dealing with objects having a mass, and within limits of classical physics, it can be shown that it is better to say that "In any interaction, to every action associated with a change of energy there is always a reaction; the associated action and reaction energy are equal in magnitude and have opposite sign. Consequently, the changes of energy contributions / forms of a single object or of more than one object interacting with each other are related and cancel out, such that the value of the total energy is not changing. So, **"A change in the amount of energy in an energy contribution / form leads to an opposite change in another energy contribution / form; this is colloquially called conversion of energy. It describes how changes of motion and position are related, e.g. described by v and h."**

What is energy not? Relation to everyday experience

Now, with the previous discussion in mind, where do problems originate? The following is a list of ideas about energy, collected by Marisa Michelini and Alberto Stefanel from other research publications (Dawson-Tunik 2004, Trumper 1993, Nicholls and Ogborn 1993). It is summarized and discussed in their contribution "Approaches and learning problems in energy teaching/learning: an overview" (Michelini and Stefanel 2010) for the workshop "Teaching about energy. Which concepts should be taught at which educational level?" held in Reims 2010.

The list of ideas is grouped here in a certain way for the further analysis:

1.   energy as a force, causing things to happen

2.   energy as something that comes in different forms, that can be converted from one form to another, that can be transferred from one object (system) to another, released by a trigger

3.   energy as something that can be created, like a product of a process, but on the contrary also energy as a quantity that is conserved

4.   energy as movement, energy as an invisible fluid, as a fuel, as a dormant ingredient in an object, a property of living things like people and animals, but also of objects

The ideas follow closely the key issues of this book, listed in section 1.1.

Group 1 refers to the observation that energy is related to changes.

Group 2 summarizes the observation that energy comes in different forms, that conversion between them is possible, that transfer or transport is possible, and also storage.

Group 3 summarizes possible origins of energy, and the scientific point of view that energy has no origin but is instead conserved.

Group 4 refers to the nature of energy, simply what energy is.

It is not surprising that the ideas about energy that common people have reflect their common, every day experience. The first two groups are actually close to the scientific concept. But this changes then in the last two groups, if there is an origin of energy or if it is conserved, and the nature of energy. It is striking that any knowledge about the origins of energy forms, conversion, and conservation is missing. But this can be understood from history: we use the concept of forces when dealing with simple situations, and not energy. The concept of energy was introduced to deal with complex situations, and initially the basis in simple situations was completely unknown. Later, science discovered connections, specifically that materials consist of many particles that have kinetic and potential energy, which we do not see. But then the description usually becomes complicated due to many effects, and the complexity of the concept of energy makes it hard to understand in general. Understanding energy related to simple situations correctly, and that complex systems are based on simple situations, is however crucial.

A general understanding is the final goal of this book; it is clear that it requires some sci.-tech. background. For the moment, lets just identify the reasons why some of the common ideas about energy are wrong, using what we learned already form simple situations.

What in detail comprises common experience about energy, and why?

What is known in some diffuse way, because commonly experienced, are the energy contributions / forms, actually better things related to them. What is seen are specifically the "indicators" like velocity v for kinetic energy, height h for gravitational energy etc. and more specific their changes. What is not seen is mass m, g etc. Even for the simple pendulum or the collision of two objects an energy contribution / form itself is not seen, and consequently also not the total energy and its conservation. It is not like with force, which is directly accessible as a concept. And it is also not like with momentum; despite we see only the velocity and not its product with mass, momentum has no different contributions / forms making it easier to understand. There is no direct observation of energy, not in an energy contribution / form, and even more not as total energy. Thus, there is consequently also no direct experience of energy conservation or its conversion. What is experienced are things somehow related to it, and without a look at the detailed and quantitative connections, things cannot be understood. It is impossible to understand what energy is directly from everyday experience. Common experience is to see the conversion of energy of one energy form to another one in some diffuse way, nothing else.

From common experience of energy conversion then wrong conclusions are drawn. The conversion of energy of one energy form to another one is observed, usually incomplete, and often resulting at the end in an energy form that is not seen anymore. For example, the ball that falls and finally rests on the floor: gravitational and kinetic energy are zero and it seems that the energy is lost. But this is wrong; it has just been converted to another energy form that is invisible (thermal energy). Then, the wrong conclusion is that energy consumption or loss would be the case in ordinary life. However, the arguments of Galilei on friction during free fall of objects allow to explain these things at least for some cases.

In general, there is a broad basis of experience what energy is related to, and what it is useful for. It has to be supplemented however with additional information and brought into the right perspective.

What can be understood from common experience is that energy enables to achieve a change, e.g. in h, v (it is not an ability). Energy in an energy form enables to make an intended change elsewhere, even to determine changes. This is by conversion to another energy contribution / form, which is associated with the desired change. E.g. when trying to get a high velocity on a bike, we can bike down a hill using the conversion of gravitational energy associated with height to kinetic energy associated with velocity. And then, the high velocity can be used to get uphill again. It is like with a skateboard or a snowboard in a halfpipe. Or like with the pendulum. And in addition, the pendulum also shows that energy is conserved by repeated conversion. Energy conversion means the amount of energy in one energy form increases, while the amount of energy in another energy form decreases. And energy conservation means simply that the changes of energy in the energy contributions / forms are pairwise with opposite sign. This can be understood from Newton's 3rd law and thus by interactions between objects. Mathematically, the amount of energy is constant, but more precise the values of the changes in the different energy contributions / forms cancel out.

For a good understanding, it is also necessary to confront with existing misconceptions directly, and remove them. Here the most crucial are addressed:

Energy is not a substance

Energy is no substance, fluid, etc. This idea came up to explain the transfer of energy, e.g. in heat transfer, and to explain the conservation of energy. But it is wrong and not needed! The pendulum shows that nothing like a substance is converted, and when two of them interact by collision also that nothing flows or is transferred mystically. Because it is common knowledge that materials consist of many particles, their kinetic and potential energy explains the true nature of energy.

Energy is not inherent in an object, e.g. released by a trigger

Energy is not "inherent" in an object. Gravitational energy does not "belong" to an object due to its mass; instead, it is due to its position in relation to another mass and the gravitational interaction between them. In common life the other mass is the earth. In $E_g = m \cdot g \cdot h$ its effect is in the value of g. In the same way, chemical energy is not inherent in a fuel like gasoline; oxygen is needed for combustion, thus combustion is also an interaction. The only exception is kinetic energy, but it only changes in interactions.

Energy is not created and / or destroyed

If a ball rests on a shelf, and possesses potential / gravitational energy by its location, the energy is just not visible compared to the velocity in the case of kinetic energy. And it is not created when a ball falls down from a shelf. From common experience it is known that it falls and thus assumed that there is energy present when the ball is on the shelf. And the same about the energy "in" human activities, fuels, or anything similar. But it is not the energy why the ball falls, it is gravitation, described by its associated force!

Another cause for confusion and wrong concepts are wrong or misleading words. E.g. commonly one energy form is preferred, e.g. kinetic energy, and then speaking of an energy loss when it decreases, e.g. due to friction. But energy is not lost, wasted, gained ...; lost, wasted, gained ... is energy of the special energy form in the focus, e.g. potential energy, kinetic energy, and fuel, gasoline ... (which carry energy) can be consumed. Science and engineering started from common experience and therefore initially made the same mistakes. This has introduced wrong and misleading wording, and as it is widespread in literature and is still used today, confuses everybody. Another example is to say in some misleading way "flow of energy", and "conversion of energy" of one energy form to another one; but energy is not a substance (fluid ...) that flows or that is changed. These misconceptions originate largely from the introduction of heat and heat flow.

Now, what could a **colloquial definition of "energy"** be? **"Energy is a concept to describe the effect of interactions between things. It has different energy contributions / forms, e.g. $\frac{1}{2} \cdot m \cdot v^2$ for the motion of an object, also called kinetic energy, or $m \cdot g \cdot h$ for the position of an object interacting by gravitation with the earth, also called a potential energy. Others are e.g. deformation, electric, magnetic, thermal, chemical, nuclear energy. A change in the amount of energy in one energy form leads to a change in another energy form; this is colloquially called conversion of energy, and describes how changes of different values, e.g v and h, are related. The sum of the energy in all energy forms is constant, which is called conservation of energy. When saying that energy is consumed, what is meant is that the amount of energy in an energy form changes, sometimes to an invisible energy form.** Assuming that common people know that materials consist of many paticles, a comment on the microscopic origin of some of the energy forms can be added.

2.1.7 Summary and conclusions

Chapter 2 discusses simple situations, which is 1 object in motion, and cases of interaction with a 2nd object. Section 2.1 "Basic terms, concepts, and laws" discussed forces in general and specifically the force of gravitation, Newton's laws of motion, momentum and its conservation, energy and its conservation, energy contributions / forms and conversion between them, work, and briefly energy storage and transport. For this, the discussion was limited to point-like objects, and the energy contributions / forms of kinetic and gravitational energy. Corresponding situations can be easily observed experimentally as well as in everyday life. They are also close to the historic way of the development of the basic terms, concepts, and laws, and allow discussing common misconceptions that occurred in history and still today, and how they are treated and corrected in the terms, concepts, and laws. This is especially the case with respect to energy and its conservation. Also, all formulas are simple; integrals can be replaced easily by products, and differentials by finite differences. Thus, up to this point, the discussion covered energy in a way that can be used to teach the basics of energy on a lower education level.

The stage of the discussion is set by the common, fundamental concepts that we have about the world

- space,

- time,

and in it

- objects (with extension, surface ...), with

- interactions between objects, affecting the motion and position of these objects

What are the **key findings**?

Looking at the terms, concepts, and laws that were developed, we can find out what we are commonly interested in; it is what they were developed for.

What we are interested in, in simple situations, is ways to describe (explain) the effect of interactions between objects on the motion and position of the objects, or changes thereof (observation), and corresponding for complex situations, as we will see in chapter 3.

If there is no interaction, nothing changes; thus, at **the core of everything are interactions**. The term **interaction** generally means "act on each other" (Hornby 1983), and is used in science, technology, as well as common life. An interaction is always between two objects. The interactions discussed so far are between an object with mass and another object with mass by gravitation, or unspecified by contact with a hand, shelf, table, and in collisions. An interaction affects the interacting objects. What we are interested in, in simple situations, is ways to describe (explain) the effect of interactions between objects on the motion and position of the objects, or changes thereof (observation), and corresponding for complex situations, as we will see in chapter 3. An interaction can affect an objects relative motion and position. An interaction, acting always between two objects, consequently affects both objects (act on each other), thus comprises an action and a reaction. Throughout history, the concepts of "force", "momentum", and "energy" were developed to describe the effect of interactions between objects.

The most intuitive concept, closest to common experience, is that of a force. The origin of forces are interactions; they describe the effect of interactions. Looking at a single object, Newton's 2nd law connects the force acting on an object with the change of its motion (Fig. 2-16). The force acting on an object always comes from interaction with a second object. Newton's 3rd law states that it is subject to a force of same magnitude but opposite direction.

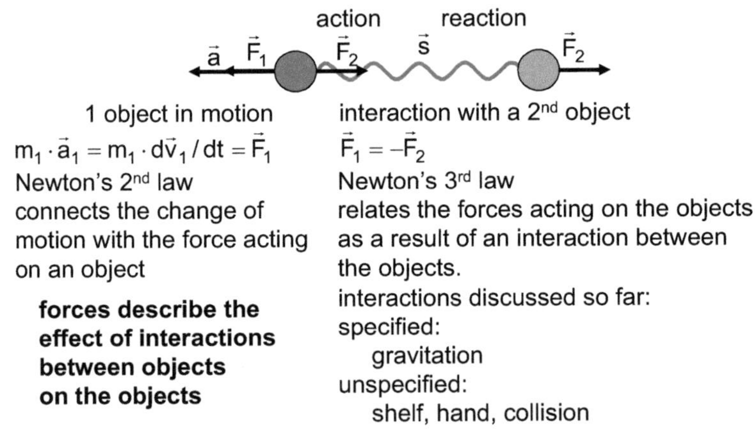

Fig. 2-16 Summary of the essentials of Newton's laws of motion, and types of interactions (often depicted like a spring) discussed so far

The effect of an interaction on motion and position can be expressed in any of the concepts of force, momentum, or energy. Fig. 2-17 shows how Newton's laws are expressed by them, related to the familiar concept of forces.

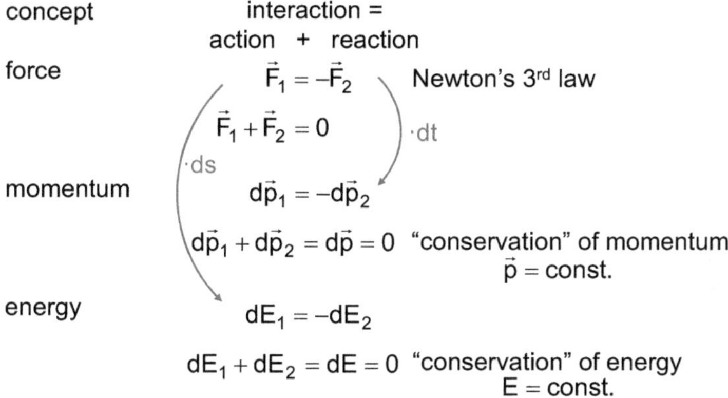

$$m_1 \cdot \vec{a}_1 = m_1 \cdot d\vec{v}_1 / dt = \vec{F}_1 \qquad \vec{F}_1 = -\vec{F}_2$$

$$dp_1 / dt = \vec{F}_1 \qquad dp_1 / dt = \vec{F}_1 = -\vec{F}_2 = -dp_2 / dt$$

$$dE_{1,kin} / ds = \vec{F}_1 \qquad \vec{F}_1 = -\vec{F}_2 = -dE_{pot} / ds$$

Fig. 2-17 Concepts of momentum and energy in their relation to the concept of force

The interaction between objects, affecting the objects relative motion and position, is the basis and thus common core in the concepts of "force", "momentum", and "energy". In all, the balance between action and reaction shows up, and if rewritten, it results in the conservation laws (Fig. 2-18).

concept   interaction =
action + reaction

force $\vec{F}_1 = -\vec{F}_2$ Newton's 3rd law

$$\vec{F}_1 + \vec{F}_2 = 0 \qquad \cdot dt$$

$$\cdot ds$$

momentum $d\vec{p}_1 = -d\vec{p}_2$

$$d\vec{p}_1 + d\vec{p}_2 = d\vec{p} = 0 \quad \text{"conservation" of momentum}$$
$$\vec{p} = \text{const.}$$

energy $dE_1 = -dE_2$

$$dE_1 + dE_2 = dE = 0 \quad \text{"conservation" of energy}$$
$$E = \text{const.}$$

Fig. 2-18 Consequences of action and reaction and in all the concepts

While having the same goal and basis, the concepts are not fully equivalent. Momentum is a vector and thus gives information on different directions, but it is only useful in dynamic situations. Energy is a scalar and thus has no more information on direction, but is useful in static and dynamic situations.

The most detailed information is given by forces, and forces are also what is related directly to our common experience. For this, energy exchanged by a force acting a distance has even a special name: work. It signifies the importance of work to achieve changes, related to a change in energy, and also the use of an energy change to do work. So why use energy? The crucial advantages of energy will become clear in the next sections.

The goal of section 2.1 is to introduce the basic terms, concepts, and laws. So, regarding the key issues stated in section 1.1, we can already conclude for simple situations the following:

1. The concept of energy describes the effect of interactions between objects on the objects. Specifically, energy correlates changes in the position and motion of objects, e.g. velocity and height, which are due to interactions. This is central to science and engineering! It is the basis to understand, optimize, and realize changes. Key to the concept of energy are energy conservation, energy contributions / forms and conversion between them.

2. Energy conservation is understood by looking at single interactions, realizing that changes are pairwise with opposite sign such that they cancel out (Fig. 2-18), as expressed for forces by Newton's 3rd law. The concept of energy describes the effect of interactions between objects; the way objects interact leads to energy changes cancelling out. Thus, strictly speaking, there is no "thing" that is conserved.

3. Energy conversion is due to a "connection" between the energy forms. The basis for conversion is e.g. the mass of an object, which connects its kinetic energy from its motion with its gravitational energy from its relative position.

To keep things simple, section 2.1 was based mainly on linear motion, and only one interaction and related force, that of gravitation. This is what we are most familiar with from common experience. These limitations are now removed, while still staying with simple situations with only 1 or 2 objects. Section 2.2 "Fundamental energy contributions" will introduce and discuss all 4 fundamental interactions and related forces, and effects due to an extension of the objects are taken into account. It will lay the foundation to discuss complex situations, having many objects and more than one interaction in chapter 3, where we will understand the full complexity of energy.

2.2 Fundamental energy contributions

In section 2.1, the behavior of 1 object in motion, and cases of interaction with a 2nd object were investigated. The objects were considered point-like, meaning effects of extension were not taken into account. Effects connected with their position were thus limited to their location, excluding orientation, and regarding motion to translation, excluding rotation. The discussion of forces covered general aspects, like Newton's laws, but regarding specifics was limited to the force of gravitation. Energy of motion, called kinetic energy, and energy of position, called potential energy, were then discussed. Within these limits, kinetic energy for translation (no rotation) is $E = \frac{1}{2} \cdot m \cdot v^2$, and potential energy as gravitational energy was simplified by $E = m \cdot g \cdot h$. This was sufficient to introduce basic terms, concepts and laws. Due to the different limitations, the discussion however also covered only a limited range of cases. The limitations are therefore now removed stepwise. In section 2.2, two limitations will be removed.

Section 2.2 covers also what happens when objects are not point-like but instead have an extension. Consequently, besides location and translation, now orientation and rotation can play a role; they lead to new energy contributions: kinetic energy of rotation, and potential energy of orientation. And besides gravitation and its associated force and energy, the other fundamental interactions are also discussed. However, the discussion will still be limited to simple situations with only 1 or 2 objects. This last restriction is necessary to stay with situations that are easy to comprehend. It will be removed in chapter 3, where we discuss complex situations.

Objects can affect each other by **4 fundamental interactions**. These are gravitation, electromagnetism, and weak and strong nuclear interaction. Each can be described by a force, to describe how an interaction affects interacting objects. The four fundamental interactions are the basis of all interactions that occur, within the atomic nuclei, atoms, molecules, stars, the solar system …. No additional interactions are necessary for a general description and understanding. Section 2.2 thus comprises and discusses all **"fundamental energy contributions"**: kinetic energy, and the potential energies of the fundamental interactions, including extension of objects.

As introduction, let's start with the key properties of the fundamental interactions and their associated forces, as well as crucial differences.

All objects having a mass interact by **gravitation**. It is always attractive, never repulsive. Consequently, its effect adds up such that gravitation typically dominates the interaction of macroscopic objects and over macroscopic distances. We experience gravitation permanently in everyday life. This is why it was discovered and investigated quite early in history, and is the reason why gravitation was selected for the discussion in section 2.1.

The **electromagnetic interaction** acts between objects having an electric charge, always also having mass. An electric charge can be positive or negative. The electromagnetic interaction is usually subdivided into a static case and a dynamic case; both were initially thought to be independent. When charged particles are at rest, the resulting force is called electrostatic force, or short electric force. When charged particles move, an additional force results, which is called magnetic force. The existence of positive and negative charges results in attractive as well as repulsive forces. This is in contrast to gravitation, and leads to a very different behavior, including the existence of dipoles and energy associated with orientation. On a microscopic scale the electromagnetic interaction is crucial; it is what keeps atoms and molecules together in materials. On larger scales, things usually have an equal number of positive and negative charges, such that the electric interaction is usually not experienced directly between larger objects; an exception is for example fluff sticking to clothes. For the dynamic case this is somewhat different, as the interaction between the earth magnetic field and the compass needle shows.

The **nuclear interaction** actually comprises two interactions. Initially discovered were the strong and the weak nuclear interaction, describing the observed interactions between the protons and neutrons forming the nuclei, called nucleons. The strong nuclear interaction explained the binding between nucleons in nuclei, the weak nuclear interaction explained radioactive decay. More detailed investigations showed later that the strong nuclear interaction actually acts between the quarks that make up the protons and neutrons and is due to the quarks having what is called color (charge). And in classifications of the fundamental interactions, the weak nuclear interaction is commonly combined with the electromagnetic interaction.

In this section, gravitation is discussed in more detail, including the introduction of fields. It is discussed where the magnetic interaction comes from, and why the nuclear interaction is not discussed deeply.

2.2.1 Kinetic energy

2.2.1.1 Kinetic energy – change of location by translation

In section 2.1 we discussed the kinetic energy of objects, meaning energy associated with motion, for the case of a translation where Eq. 2-26 holds. It is then specifically called **translational kinetic energy** $E_{kin,\,trans}$

$$E_{kin,trans} = \frac{1}{2} \cdot m \cdot v^2 .$$

Eq. 2-32

The object we discussed was a ball, to keep things simple. Similar examples, very common in everyday life, are the energy of motion of a driving car or train, or a flying plane. Also included is the energy of motion of a mass of water flowing in a pipe or in a river, of the air moving when the wind blows, thus cases without an object described by a well-defined surface and mass. And of course, on the microscopic scale, included is the energy of motion of electrons etc.

2.2.1.2 Kinetic energy – change of orientation by rotation

Besides translation, the motion of objects can also be a rotation. Associated with it is rotational kinetic energy. Fig. 2-19 shows the rotation of a wheel.

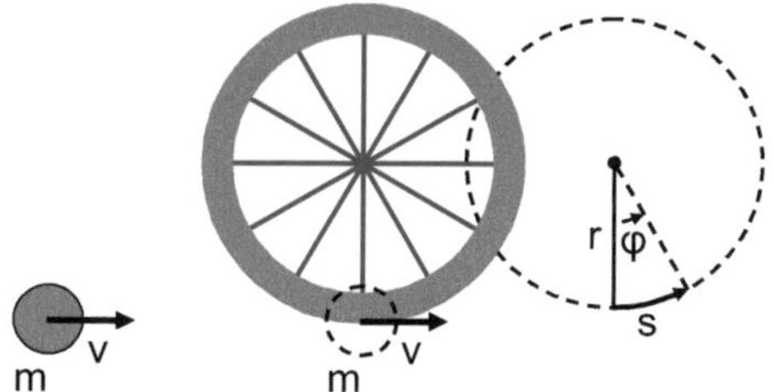

Fig. 2-19 Connection between translational and rotational kinetic energy

First, let's just focus on a part of the wheel, as marked in Fig. 2-19. For a short distance it moves straight such that Eq. 2-32 applies. But generally the motion is circular, with $s = r\cdot\varphi$, and consequently $v = r\cdot\omega$ (section 1.2). The latter must therefore be inserted in Eq. 2-32 to get the correct kinetic energy of the part of the wheel taking into account that it rotates. This gives

$$E_{kin,trans} = \frac{1}{2} \cdot m \cdot r^2 \cdot \omega^2 .$$

Eq. 2-33

The rotational kinetic energy of the wheel is the sum of the kinetic energy of all its parts, denoted by i. As they have the same angular velocity ω, it is

$$E_{kin,rot} = \sum_i \frac{1}{2} \cdot m_i \cdot r_i^2 \cdot \omega_i^2 = \frac{1}{2} \cdot \left(\sum_i m_i \cdot r_i^2 \right) \cdot \omega^2 .$$

Eq. 2-34

The **rotational kinetic energy** $E_{kin,\,rot}$ can thus be written as

$$E_{kin,rot} = \frac{1}{2} \cdot \Theta \cdot \omega^2 ,$$

Eq. 2-35

similar to the translational kinetic energy in Eq. 2-32. The term

$$\Theta = \sum_i m_i \cdot r_i^2$$

Eq. 2-36

is the **rotational inertia**, often also called **moment of inertia** because of its mathematical basis in Eq. 2-36. If an object moves free, e.g. a thrown ball, it has a point that moves smoothly, called its **center of mass** (CM), while its rotation is around it. For a ball or the earth, the CM is at its center. Fig. 2-20 shows the case of an object composed of two equal masses being connected.

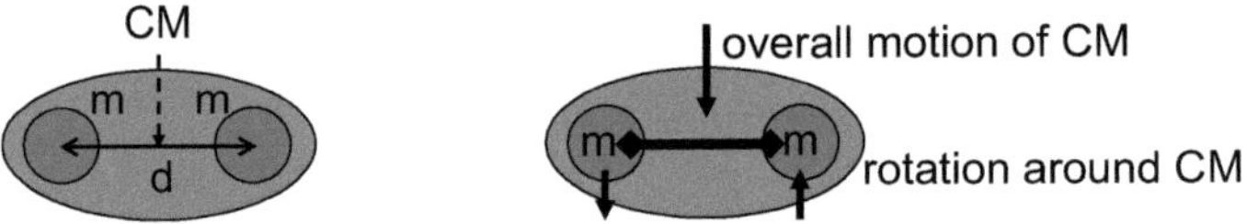

Fig. 2-20 Object with a simple mass distribution in motion

The rotational kinetic energy is not new, but just a convenient description of what in detail can be seen as many small linear translations. Thus, translational and rotational motion share many similarities. For example, the velocity v corresponds to the angular velocity ω, and the mass m to the rotational inertia Θ. Rotation is also associated with a momentum, given by $\Theta \cdot \omega$ and called **angular momentum**; it is also conserved. There is also a rotational impulse, and the torque τ replaces the force F in Newton's 2nd law. Because this book is specifically about energy, and linear momentum was mainly introduced as an aid to understand energy and its conservation, we will not discuss details here; they can be found e.g. in Halliday et al. 1993.

2.2.2 Gravitational energy

2.2.2.1 Where does gravitation come from?

Gravitation acts between all objects having a mass m. It is always attractive, never repulsive. As a result, gravitation typically dominates the interaction of macroscopic objects and over macroscopic distances. We experience gravitation permanently in our life. This is why it was discovered and investigated early in history, already Aristotle discussed it, and it is the reason why the force of gravitation was part of the discussion in section 2.1.

Up to this point, we discussed gravitational energy of an object just as a consequence of the force of gravitation $F_g = m \cdot g$ as $E_g = m \cdot g \cdot h$. We did not consider the source of the force, which is the interaction by gravitation with a second object. On earth, all things fall to the ground, so towards the earth. Thus, for objects on earth the earth must be the second object and somehow enter the equation in the value of g. But how?

2.2.2.2 General formulation of the force of gravitation

The discovery of the law of gravitation, describing the force, took a long time. Aristotle started, thinking that objects with different mass fall at a different rate, e.g. feathers and stones; this is just what is common experience. Galilei then showed that Aristotle was wrong; neglecting the friction of air, all objects accelerate towards the earth at the same rate. For the next step, new observations and data were needed; they did not come from the earth, where g is the same everywhere, but from the sky. In 1597, Johannes Kepler (1571 to 1630) suggested that the sun affects the planet motions by a force. He then used data on the planet motions, from observations by Tycho Brahe (1546 to 1601), to develop mathematical laws, called **Kepler's three laws**, which describe the motion of all planets in the same way. But they do not yet describe the force that acts. Newton then found a mathematical description of the force, which is the force of gravitation. It describes the interaction between masses by gravitation, which is attractive. For two objects with mass m and m_x (Fig. 2-21, left) the force of gravitation F_g is, according **Newton's law of gravitation**, which he published in 1687

$$F_g = m \cdot m_x \cdot \frac{1}{r^2} \cdot G \, .$$
Eq. 2-37

G is the **gravitational constant**, with $G = 6.67 \cdot 10^{-11}$ N·m^2/kg^2, and r is the distance between the masses (considered to be point-like). In other words, r is the relative position, specifically the relative location.

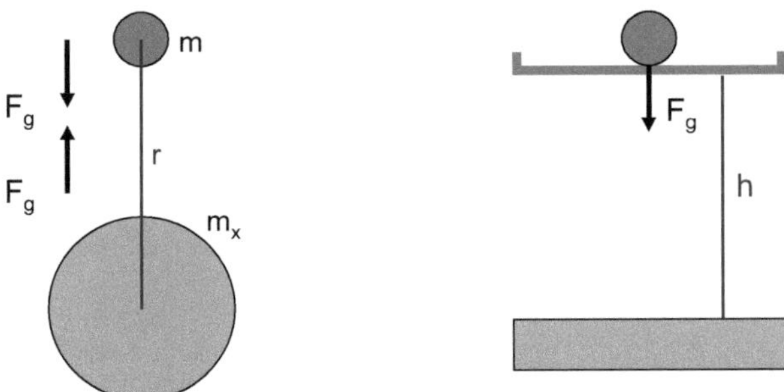

Fig. 2-21 Force of gravitation: general (left) and simplified case (right)

According to Newton's 3rd law, if a force acts on an object there is another force of equal size and opposite direction, and it is due to another object. For somebody on the earth, it is the earth with its mass. But also the moon goes around the earth, and the earth itself goes around the sun. It is like swinging a mass on a rope in a circle; it is necessary to pull on the rope to force the mass moving in a circle. For the planets it is the force of gravitation by interaction with the sun, and for the moons it is the planets; this was a crucial discovery. Newton could explain Kepler's results by his laws of motion, and especially derive Kepler's 3rd law using his law of gravitation (derivations for all in Halliday et al. 1993). Gravitation was the first interaction to be described mathematically; even more, it demonstrated that the universe is governed by the same laws everywhere (Rooney 2011); the same law describes the fall of an apple on earth as well as the motion of the planets around the sun, and the moons around the planets.

Rearranging Eq. 2-37 allows separating the objects relevant property, its mass m, from the rest, which is now enclosed in brackets

$$F_g = m \cdot \left(m_x \cdot \frac{1}{r^2} \cdot G \right).$$

Eq. 2-38

This gives a simplified description of the force using the term **(force) field**, an approach frequently used in physics. To calculate the force it is sufficient to know the field (the value is in the brackets) at the location of the mass m; it is not necessary to know the detailed origin of the field, for example if it is due to several masses m_x, or where they are located. The force of gravitation on the surface of the earth is due to the interaction with the mass of the earth, and it is practically the same anywhere at the surface as r is the same. Therefore, the force of gravitation at the earth surface can be described by

$$F_g = m \cdot g \, ; \qquad\qquad \text{Eq. 2-39}$$

g = 9.81 N/kg, often approximated by a value of 10 N/kg. Thus, g represents the force field by gravitation at the earth surface, and says that on a mass of 1kg a force of 10N acts.

2.2.2.3 Gravitational energy with respect to location

What about the gravitational energy of two objects interacting by gravitation? This will also reveal some important aspects about energy related to interactions in general. Afterwards, we will look again at the simple case at the earth surface.

We do not know anything about the interaction, since it is a fundamental interaction, except from how it affects objects. Thus, we need to apply the work - energy theorem (Eq. 2-23): the change of the gravitational energy is equal to the work done when changing the distance between both objects

$$\Delta E_g = W = \int_i^f \vec{F}_g \cdot d\vec{s} \, . \qquad\qquad \text{Eq. 2-40}$$

From the force of gravitation between two (point) masses (Eq. 2-38) follows for their **gravitational energy** E_g (ds parallel to F, thus becoming dr) that

$$E_g = \int_\infty^r \left(m \cdot \left(m_x \cdot \frac{1}{r^2} \cdot G \right) \right) \cdot dr$$

$$= m \cdot m_x \cdot G \cdot \int_\infty^r \frac{1}{r^2} \cdot dr = m \cdot m_x \cdot G \cdot \left| -\frac{1}{r} \right|_\infty^r$$

$$= -m \cdot m_x \cdot G \cdot \frac{1}{r} \, . \qquad\qquad \text{Eq. 2-41}$$

Gravitational energy is potential energy. The zero point is chosen at $r = \infty$. This is not an arbitrary choice. It is common to choose the zero point of energy where the force is zero, meaning where the interaction is negligible. For the case of kinetic energy, when the velocity of an object is zero it is not capable to do work on another object; and the same holds if an interaction is negligible. Similarly, the zero point of energy of your car is when the tank is empty. Energy is about changes most of the time, but there are limits.

Because the force of gravitation is always attractive, the gravitational energy becomes smaller when the masses approach each other, and as it is normalized to zero at $r = \infty$ it becomes negative, as shown in Fig. 2-22.

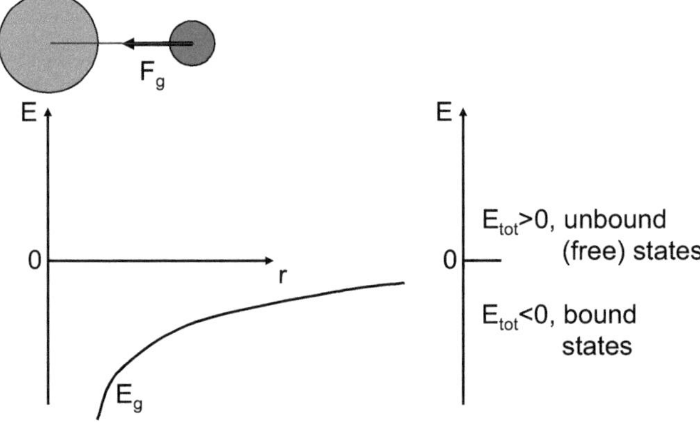

Fig. 2-22 Gravitational energy of two objects

An attractive interaction has the consequence that work has to be done to increase the distance between the interacting objects, and this is associated with an increase of the interaction energy. If this is the case in all relevant directions, the situation is to some extent stable against disturbances from outside, and the objects are said to be **bound** to each other.

Let's now discuss the simplified and common case of an object at the earth surface, that was already discussed in section 2.1 and 2.2. For small changes of height above the surface of the earth, the field of gravitation is homogeneous and the resulting force on a mass is constant. The distance of relocation is the change in height. Then, the associated **gravitational energy** E_g is

$$E_g = m \cdot g \cdot h .$$

Eq. 2-42

This is the formula discussed before in section 2.1; it is what is directly observed at the earth surface. Examples are a book on a shelf or a stone on top of a wall. But the simple formula often leads to wrong conclusions. This solution, missing the origin of the field and thus force, in this case the earth, leads to the common idea that the gravitational energy belongs to the object lifted from the ground. But the gravitational energy does not belong to the object that is lifted! As an interaction energy it belongs to the interaction between both objects, the object that is lifted and the earth, and is dependent on their distance, which we experience as height above the ground; we only see a small part of the whole situation. The effect of the second, the earth, is "hidden" in the value of g. Second, the free choice of the zero point of energy in Eq. 2-42 does not imply that only energy differences are relevant; the general case shows that there is a zero point.

Now, what if the object is not just lifted, but maybe also moved sideways? In section 2.1.4.3 we already discussed that a force acting perpendicular to the direction of relocation has no effect on energy; and no work is done. Now, how are work and energy affected if the force is not parallel and not perpendicular to the direction of relocation? As stated by Eq. 2-30, the effective part of the force is the part of the force resulting from its projection on the direction of relocation.

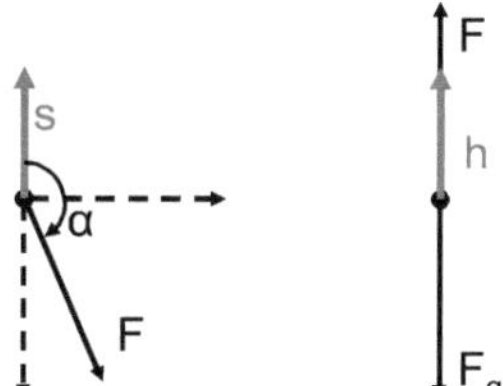

Fig. 2-23 Scalar product of the force and distance of relocation

Thus, in general, if a force F acts a distance s (Fig. 2-23, left) the associated work and thus change in energy is given by the **scalar product**

$$\vec{F} \circ \vec{s} = \left|\vec{F}\right| \cdot \left|\vec{s}\right| \cdot \cos \alpha , \qquad \text{Eq. 2-43}$$

where F is the acting force, s is the distance of relocation, and α the angle between them. Often, "$\circ$" is used to show that a product is a scalar product.

For example, what if a mass m is pulled up the slope of a hill? Neglecting friction, the only force that acts is the force of gravitation. Thus, for the work done and the energy difference between the initial and the final state it is the high that counts and not the length of the slope. The slope increases the distance the mass has to be pulled, but the same way the force needed to pull against the force of gravitation is reduced. This could lead to the idea that it is easiest to use a slope with an inclination as small as possible. But if there is friction, the increasing length of the slope would increase the energy needed to pull the mass up the hill.

2.2.2.4 Gravitational energy with respect to orientation

The discussion of kinetic energy showed that the extension of an object can lead to a new energy contribution: rotational kinetic energy. Strictly speaking it was not new; Eq. 2-35 is a better way of describing the coordinated motion of the mass distributed in a rigid object. What about gravitation?

A simple example of an object with an extension and thus a mass distribution within is an object consisting of two equal masses m at a distance d, as shown in Fig. 2-24.

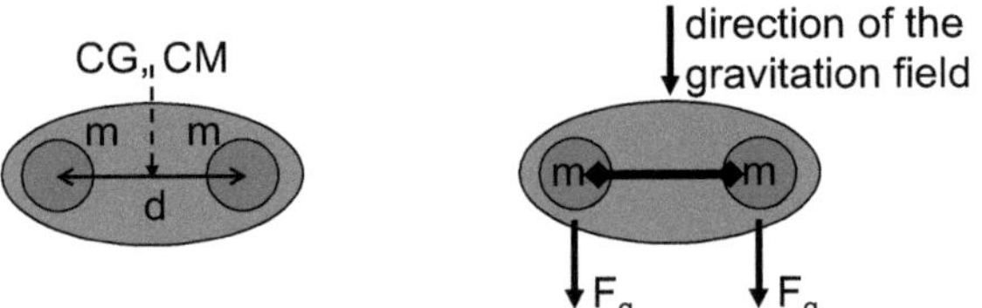

Fig. 2-24 Object with a simple mass distribution in a gravitation field

The CM has already been used to split translational and rotational energy, so let's use it again. An object has gravitational energy by the location of its CM in a gravitation field. And its orientation? In a gravitation field there is a point in an object, called the **center of gravity** (CG), where the object experiences no torque and thus does not start to rotate if it is supported there. Thus, if CG and CM are the same there is no effect of orientation. This is the case in a homogeneous gravitational field, meaning where g is the same everywhere. Thus, the energy of an object in a homogeneous gravitational field, even if it has an extension, is described just by the location of its CM.

2.2.3 Electromagnetic energy

2.2.3.1 Where does the electromagnetic interaction come from?

The **electromagnetic interaction** acts between objects having an electric charge q, always being in connection also with mass and thus gravitation. An electric charge can be positive or negative, leading to forces that can be attractive as well as repulsive. This is in contrast to gravitation, and leads to a very different behavior.

The electromagnetic interaction is usually subdivided into a static case and a dynamic case. When objects with electric charge are at rest, the resulting force is called **electrostatic force**, or short **electric force**. When they move an additional force results, called **magnetic force**. It is often practical to deal with both separately as both have very different behavior, and actually, because of that they were even discovered separately.

In nature, usually all except very small objects have an equal number of positive and negative charges, and consequently no net charge. Thus, for the static case, there is usually little or no interaction between two large objects, or a large and a small object. Consequently, between large objects usually gravitation dominates such that the electric interaction is commonly not experienced directly in everyday life; an exception is for example charged fluff sticking to clothes. For the dynamic case it is somewhat different as the interaction between the earth magnetic field and the compass needle shows. On a microscopic scale it is the electromagnetic interaction that keeps atoms and molecules together in materials (chapter 3).

2.2.3.2 General formulation of the electric force

Already in ancient times people knew that rubbing amber creates a force that makes it attract straw, but the nature of the force was entirely unclear. Thomas Browne (1605 to 1682) gave the effect the name "electric", because the Greek word for amber is "electron" (Halliday et al. 1993). According to Rooney 2011, the first machine that was able to generate the effect, called electrostatic generator, was built in 1603 by Otto von Guericke. In 1744 then, the first device to store electricity, the Leyden jar (a capacitor), was developed. Generator and storage devices allowed better experiments, making it possible to clarify the nature and properties of the force.

That electricity might have positive and negative charges was suggested by Benjamin Franklin (1706 to 1790). C. A. Coulomb then derived the law describing the force between electric charges at rest from experiments done in 1785 (Halliday et al. 1993). This was almost 100 years after Newton published the law of gravitation. The force between two objects of electric charge q and q_x at rest (Fig. 2-25), called electrostatic or just electric force, and also **Coulomb force**, is described by **Coulomb's law**

$$F_{el} = q \cdot q_x \cdot \frac{1}{r^2} \cdot \frac{1}{4\pi \cdot \varepsilon_0} = q \cdot \left(q_x \cdot \frac{1}{r^2} \cdot \frac{1}{4\pi \cdot \varepsilon_0} \right). \qquad \text{Eq. 2-44}$$

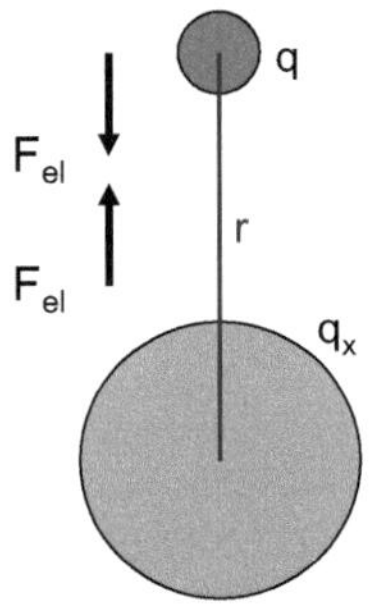

Fig. 2-25 Electric force between charges, also called Coulomb force

Coulomb's law is very similar to Newton's law of gravitation (Eq. 2-37). There is again a universal constant, the **permittivity constant in vacuum** ε_0, and r is again the distance, now between the point-like electric charges. There is however a significant difference: while the force of gravitation is always attractive, as mass is always positive, electric charges can be positive (+) or negative (-), and the electric force between two charges is attractive for charges of opposite sign and repulsive if they have the same sign.

As for gravitation, the description can be simplified using the field concept. The electric force on a charge q located in an electric (force) field $\vec{E}$ is

$$\vec{F}_{el} = q \cdot \vec{E}, \qquad \text{Eq. 2-45}$$

where the field is given by the term in brackets in Eq. 2-44. The direction of the field is originating from a positive charge towards a negative charge, and the electric field in total can be due to several electric charges q_x.

2.2.3.3 Electric energy with respect to location

How is the electric force related to electric energy? Similar as for gravitation, the electric energy E_{el} is due to the electric interaction and a change in the distance between charges. As the laws describing the force of gravitation (Eq. 2-37) and the electric force (Eq. 2-44) are so similar, a derivation is quite easy. Using the analogy to Eq. 2-41, the **electric energy** E_{el} of two (point) charges at a distance r is

$$
\begin{aligned}
E_{el} &= \int_{\infty}^{r} \left(q \cdot \left(q_x \cdot \frac{1}{r^2} \cdot \frac{1}{4 \cdot \pi \cdot \varepsilon_0} \right) \right) \cdot dr \\
&= -q \cdot q_x \cdot \frac{1}{4\pi\varepsilon_0} \cdot \frac{1}{r} \, .
\end{aligned}
$$

Eq. 2-46

It is only a function of position, and thus again a form of potential energy. When the distance r between the charges is infinitely large, the force is zero, and this is again used to define the zero point of electric energy (Fig. 2-26).

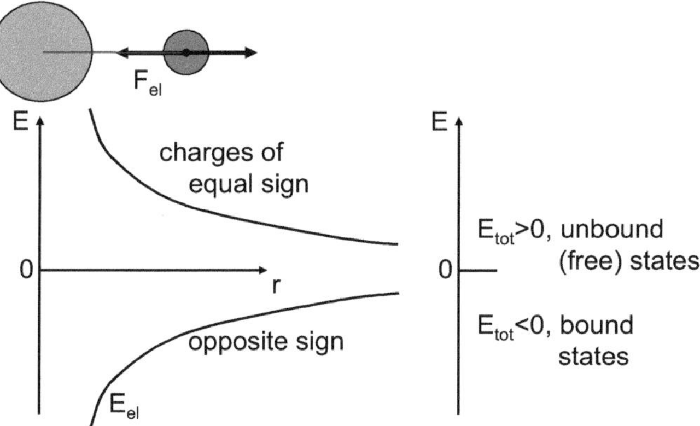

Fig. 2-26 Electric energy of two objects

For charges of opposite sign the force is attractive, and thus the energy becomes negative when the charges approach each other; for charges of equal sign the force is repulsive, and thus the energy becomes positive. In general, the attractive interaction between charges of opposite sign has the consequence that work has to be done on the system of two charges from outside to separate the charges further; this is associated with an increase of the interaction energy. If this is the case in all relevant directions, the consequence is that the system is stable against external disturbances, and the objects are said to be bound to each other. Otherwise, they are unbound, e.g. when the interaction is repulsive between charges of equal sign.

For the simplified case of a constant electric field E and a relocation by a distance s parallel to the force F, the **electric energy** E_{el} an be expressed similar to Eq. 2-42 as

$$E_{el} = \vec{F}_{el} \circ \vec{s} = q \cdot E \cdot s .$$

Eq. 2-47

The notation used here uses "E" for energy as well as for the electric field. This is common usage in literature, and is inevitable. Which one is meant is usually clear form the context, and additional notation as an arrow for the field or a subscript for the energy form should avoid ambiguities.

2.2.3.4 Electric energy with respect to orientation

Up to this point, the discussion referred to an object having a single charge. Like before for mass and gravitation, we will now discuss the case of an object with extension, specifically now a charge distribution in the object.

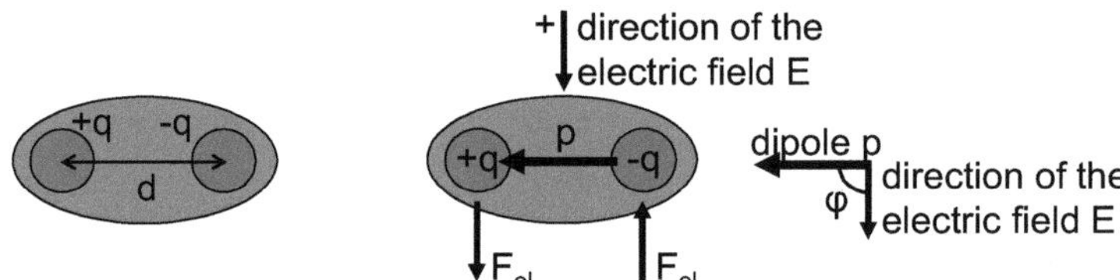

Fig. 2-27 Object with a charge distribution in an electric field

For simplicity again, and because of its practical importance, let's look at the case where the object has a charge +q and –q at a distance d (Fig. 2-27), and is exposed to a homogeneous electric field due to the presence of another charge, e.g. a positive one. The two charges are then subject to forces of the same magnitude but opposite in direction. Thus, in contrast to gravitation and mass (Fig. 2-24), in this case

- the object is now not subject to a net translational force, as due to the opposite charges the net charge is zero; thus it has no energy of location

- the two forces now create a torque, and cause energy of orientation.

66

The associated electric energy of orientation is simply the sum of the individual energies of the charges relocated a distance s or -s correspondingly

$$E_{el} = +q \cdot E \cdot s + (-)q \cdot E \cdot (-s).$$

Eq. 2-48

With the rotation around the center, $-s = d/2 \cdot \cos(\varphi)$; the "-" originates from the fact that the lowest energy is when the negative charge is oriented towards the external positive charge. Then, Eq. 2-48 becomes

$$E_{el} = -q \cdot d \cdot E \cdot \cos(\varphi).$$

Eq. 2-49

This can be rewritten in a much simpler way as follows. The two charges of equal magnitude and opposite sign at a distance d, called an **electric dipole**, are represented by a vector pointing from the negative to the positive charge (Fig. 2-27, right). Defining the **electric dipole moment** p as

$$p = q \cdot d,$$

Eq. 2-50

the **electric energy of an electric dipole** in an electric field is

$$E_{el} = -\left|\vec{p}\right| \cdot \left|\vec{E}\right| \cdot \cos \varphi = -\vec{p} \circ \vec{E}.$$

Eq. 2-51

It is a function of orientation, by the angle, and thus also a potential energy. It is zero for perpendicular orientation, the same as without electric field. The lowest energy, with negative value, is when the dipole is oriented parallel to the external electric field (Fig. 2-27, right), and the highest energy, with positive value, is for antiparallel orientation.

Electric energy thus refers to the energetic effect of the location of a (net) charge, and the orientation if for example two charges of equal magnitude and opposite sign (no net charge) form a dipole.

In general, an object can of course contain many different charges and thus form a **multipole** instead of just a **dipole**; this is however not discussed here because these do not add any new insight and we do not need them later. However, what is important is that for an inhomogeneous electric field things are somewhat different; if two dipoles, each having zero net charge, are close to each other they experience an attractive force for this reason. This force tends to align them antiparallel and next to each other. For two magnets, essentially being magnetic dipoles, this is a common experience. So let's now look at the magnetic interaction.

2.2.3.5 Historic background and observations of the magnetic force

Magnetism is "part" of the fundamental interaction of electromagnetism. Historically, two independent interactions, electric and magnetic interaction, were discovered; they were later found to be actually a single one, therefore called electromagnetism. Dealing with static charges is the electric part, which depends only on the relative location; we have just discussed it. Dealing with moving charges, meaning electric currents, is the magnetic part. This is what we will discuss now.

According to Rooney 2011, the power of some materials to attract iron was already known in ancient times, for example Aristotle reports a description by Thales (about 625 to 545 BC). The use of magnetism for navigation, employing the discovery that a magnetic needle allowed to move free always points in the same direction, is confirmed latest by 1117 in China. William Gilbert (1544 to 1603) then made first systematic investigations and in his book "De magnete", published in 1600, explained the orientation of the compass needle by the earth itself being magnetic. Thus, the magnetic needle of the compass, which is free moving, experiences a torque that moves it to its lowest energetic orientation; the compass needle then points to the North Pole (Fig. 2-28, center). This resembles the behavior of an electric dipole. The magnetic poles are called north N and south S, like + and -.

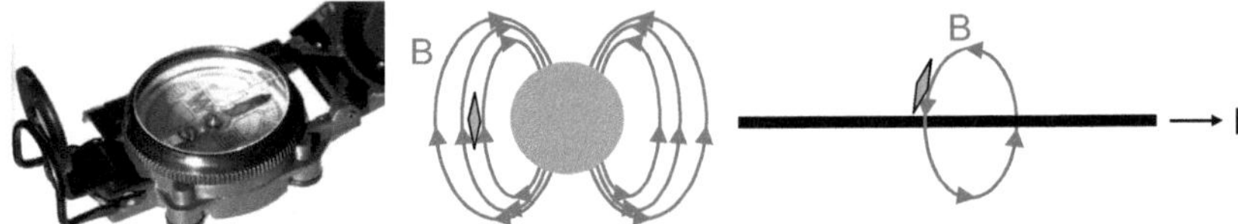

Fig. 2-28 Compass with magnetic needle (left), oriented by the magnetic field of the earth (center) as well as by an electric current (right)

In 1820, Hans Christian Ørsted (1777 to 1851) found a connection between electricity and magnetism, observing that an electric current in a wire can affect the orientation of a magnetic compass needle (Halliday et al. 1993). Thus, a force and a related magnetic field must exist around the wire with its electric current. Shortly after, André-Marie Ampère demonstrated that when parallel wires carry an electric current, they also attract or repel each other depending on the currents going in the same or opposite direction.

2.2.3.6 Theoretical background of the magnetic force

The fact that electric and magnetic interaction belong to a single interaction, called electromagnetic interaction, is explained by **relativistic mechanics**.

In classical mechanics, observations are made at rather small velocities. Then, velocities add up, e.g. the velocity of an arrow that is shot from within a car when observed from the outside is the velocity with reference to the bow plus the velocity of the bow moving with the car. But experiments have shown that this does not hold at very high velocities. Especially, observations show that the velocity of light c is independent of the velocity of the light source, meaning it is the same in all inertial reference frames. This has serious consequences on time and length in different reference frames, and is what Einstein formulated in his **special theory of relativity**, published in 1905. A nice treatment contains Tipler and Mosca 2004.

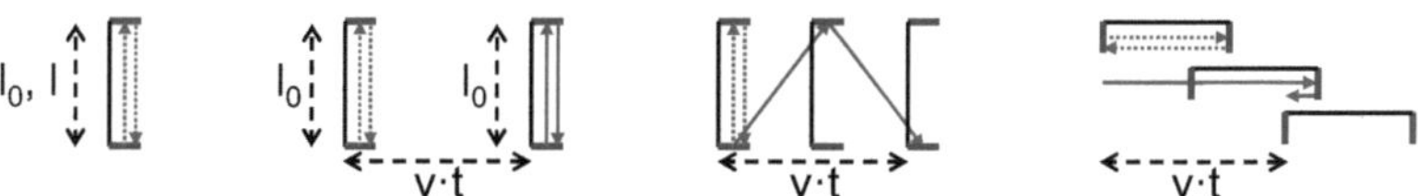

Fig. 2-29 Light travelling between mirrors in different situations, from left (1) to right (4)

Fig. 2-29 shows light travelling between two mirrors, forward, then back. The mirrors are at a distance of l_0 in their own reference frame, all the time. In the lab reference frame the length l is also l_0 if the mirrors are at rest (1). If the mirrors move in the lab reference frame, such that the light travels perpendicular to the direction of motion (2), both lengths are still the same. But since the light has to travel a longer way in the lab frame (3), while c is the same in both reference frames, the time in the lab frame t must be longer than in the mirror frame t_0. This effect is called **time dilation**, and

$$t = t_0 / \sqrt{1-\left(v/c\right)^2} \ .$$
Eq. 2-52

If the light travels parallel (4), analysis shows that the light in the lab frame travels a shorter distance than in the mirror frame. This effect is called **length contraction** (in direction of motion), and using Eq. 2-52 follows

$$l = l_0 \cdot \sqrt{1-\left(v/c\right)^2} \ .$$
Eq. 2-53

Now, how does this explain the following observations: two electrically conducting wires experience a force if, and only if, both wires conduct an electric current; the force is attractive if the currents are parallel, and repulsive if anti-parallel.

An electric current in a wire means that there are moving electric charges. Let's first look at the explanation of the case of parallel currents; it is based on an explanation by Benson 1991. As Fig. 2-30 shows, both wires are electrically neutral such that there is no electric force. In both wires, positive and negative charges move in opposite directions, here for convenience at same velocities, such that a net current results. Let's focus on a single, positive charge in the upper wire. In the reference frame of that charge, the positive charges of the other wire are at rest, while the negative ones move. Due to length contraction, the negative charges have a shorter distance and thus higher density; the wire seems to be negative charged, which results in an attractive force. In the lab frame this attractive force is observed, and interpreted as "magnetic" force. The same is true looking at the negative charges in the upper wire such that the observation is explained. The electric force in one reference frame is thus a magnetic force in another reference frame. Consequently, electric and magnetic interaction belong to a single interaction, called "electromagnetic interaction", as explained by special relativity.

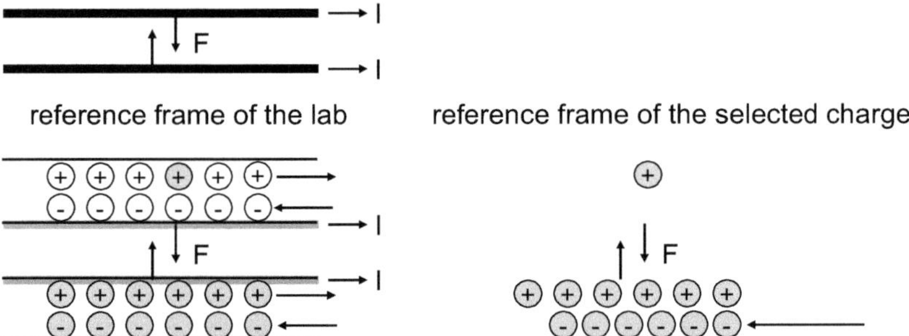

Fig. 2-30 Case of parallel electric currents; left: lab reference frame; right: reference frame of the selected charge

The other observations can be explained similarly. If the currents are anti-parallel the effect is reverse, resulting in a repulsive force. If there is no current, then there is no effect and thus no force. And if there is just one current there is no force as the effects on positive and negative charges cancel out.

2.2.3.7 Description of a magnetic force by a magnetic field

Now, how are the previously discussed observations described by a magnetic field? That an electric current creates a magnetic field was clear from the observation by Ørsted (Fig. 2-28, right), including the direction of the field. The resulting field shows again Fig. 2-31, left. The right of Fig. 2-31 shows the combination with the observation of a force between parallel currents by Ampère. The force is attractive, thus at right angle with the magnetic field as well as with the direction of motion of the electric charges. The same results if the magnetic field is instead due to a permanent magnet.

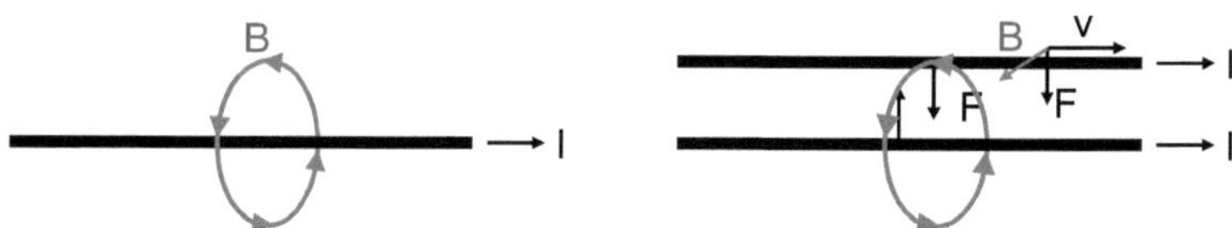

Fig. 2-31 Magnetic field due to moving electric charges, and resulting force on other moving charges

Many charges moving as an electric current e.g. in a wire cause a magnetic field, thus also a single moving electric charge. And a magnetic field causes a force on many charges moving as electric current e.g. in a wire, thus also on a single moving electric charge. The force F on a moving electric charge q due to a magnetic field B, as observed from experiments (Fig. 2-31, right), is called **Lorentz force** and described using the **vector product** by

$$\vec{F}_{mag} = q \cdot \vec{v} \times \vec{B} = q \cdot |\vec{v}| \cdot |\vec{B}| \cdot \sin(\alpha);$$

Eq. 2-54

it is perpendicular to the motion of the electric charge as well as to the magnetic field B. Since the force on the el. charge is perpendicular to its motion, $\cos(\alpha) = 0$, and consequently (section 2.1.4.3) its energy does not change! But it can force the el. charge into a circular motion, and as is discussed soon (Fig. 2-32) this has then its own magnetic field; the field of a dipole. Altogether, the electromagnetic force on an electric charge in electric and magnetic fields is from Eq. 2-45 and Eq. 2-54

$$\vec{F}_{el,mag} = q \cdot \vec{E} + q \cdot \vec{v} \times \vec{B}.$$

Eq. 2-55

Compared to the gravitational and electric force, the magnetic force is not radial; this is striking, however it is just due to being a relativistic effect.

Infinitely long and straight currents, as just discussed, do not exist in reality. A free moving charged particle gets close, but in wires they are impossible. Common in nature and technology are closed current loops e.g. in wires. They cause a magnetic field, shown in Fig. 2-32; it is the field of a dipole (the field resulting from a straight current shows already Fig. 2-31, and the dipole field resulting from a current loop can be derived from it). The explanation for the magnetic field of the earth (Fig. 2-28) is also an electric current, circulating in the interior of the earth.

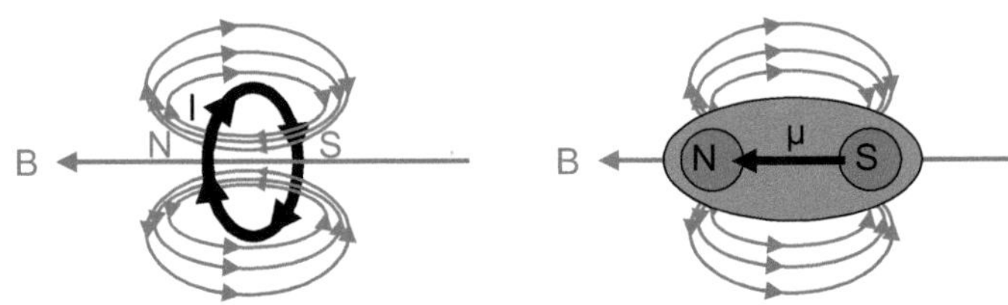

Fig. 2-32 Magnetic field B caused by an electrical current loop, left, and simplified representation by a magnetic dipole, right

The magnetic field of a magnetic dipole is similar to the electric case. It is described by a **magnetic dipole moment** μ, like a small magnet. The magnetic field by definition originates from what is labeled north, N, and goes to what is labeled south, S (Fig. 2-32). This comes from the use of magnets as compass; the N of the magnetic needle indicates where the North Pole is, meaning that at the geographic north N is the physical south S (Halliday et al. 1993), and that the magnetic field lines converge to it. The orientation of the dipole is marked by a vector from S to N, as if the magnetic field lines continue (Fig. 2-32), the same as for an electric dipole (Fig. 2-27), with N and S replacing + and -.

A magnet can be natural or due to an artificial electric current. In both cases, many of the features of magnets can be explained by current loops. For example, if a natural magnet is divided, smaller magnets with N and S poles each result, and if smaller ones are combined, a larger dipole results. The same is observed if the magnets are replaced by artificial current loops. And as is clear from Fig. 2-32, there is no **magnetic charge**.

2.2.3.8 Magnetic energy with respect to orientation

There is no magnetic charge, and thus no energy associated with the location of a magnetic charge. There are only magnetic dipoles, and the function of the compass already indicates the existence of a torque, and associated with it energy of orientation. This shows Fig. 2-33; it is similar to Fig. 2-27.

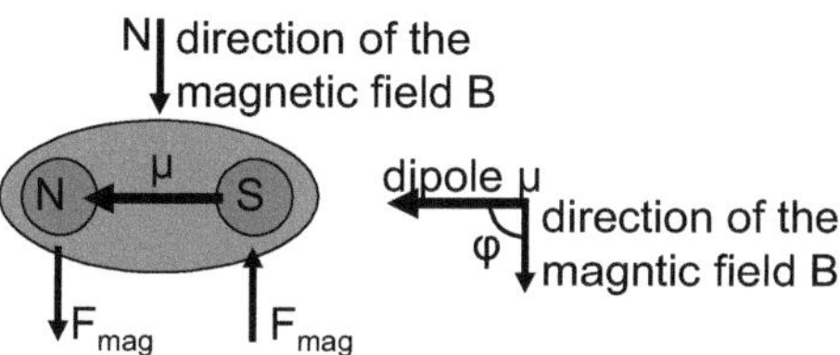

Fig. 2-33 Magnetic dipole in a magnetic field

The magnetic dipole moment μ has an orientation to the magnetic field B, and its energy depends on its orientation. Similar to the electric case, described by Eq. 2-51, the **magnetic energy of a magnetic dipole** in a magnetic field B is

$$E_{mag} = -|\vec{\mu}| \cdot |\vec{B}| \cdot \cos\varphi = -\vec{\mu} \circ \vec{B} .$$
Eq. 2-56

It is again potential energy, a function of orientation and thus angle, and it is lowest when oriented with the magnetic field, and highest when opposite.

As dipoles themselves are the origin of a magnetic field, dipoles interact with dipoles. As a result, dipoles tend to orient each other in a way that their poles are $S - N \cdots S - N \cdots S - N$ to be at the lowest energy state.

Fig. 2-34 Interaction of magnetic (left) and electric dipoles (right)

But even more they also attract each other. This is well known from magnetic dipoles. But it is easier to understand for electric dipoles: the charges that are closest cause the strongest force such that overall the dipoles first reorient, and then attract each other. This can also be understood from the respective fields, which are for dipoles inhomogeneous as Fig. 2-32 shows.

2.2.4 "Nuclear energy"

2.2.4.1 Where do the nuclear interactions come from?

The nuclear interactions are the last of the fundamental interactions not yet discussed. The topic is somewhat hard to grasp because of its development, and its complexity. Initially, the strong and the weak nuclear interaction were introduced, and the terms are still used. However, their observed behavior is complex, quite different from the other interactions. It was later found that there is a much simpler basis. But let's start from the beginning.

After the discovery in 1908 that atoms have a nucleus, a question came up: why do the protons and neutrons making up the nucleus, called **nucleons** (Fig. 1-1), stick together, despite the highly repulsive electric force between the protons that are so close in the nucleus. The answer was the introduction of the **strong nuclear interaction** with its associated force. But there is a limit: if too many protons are in a nucleus it is not stable anymore. Experiments showed that the strong nuclear interaction is extremely short range, attractive between about 1 to $2.5 \cdot 10^{-15}$m, above negligible, and below repulsive. This is quite different from the other fundamental interactions, and not the only difference. But there is no need to go into detail. Besides the question why the nucleons stick together, there were other nuclear phenomena needing an explanation, like beta decay. The answer was the **weak nuclear interaction**, discovered 1934 by Enrico Fermi. Therefore, there are two different **nuclear interactions**, the strong and the weak nuclear interaction. Both have a short range, acting only inside atomic nuclei. And what makes things quite complicated is that nucleons also have mass and electric charge, such that the nuclear interactions and associated forces are not observed alone; there is also gravitation that can however be neglected, and there is the electromagnetic interaction between the charged protons that can play a significant role, even make nuclei unstable as said above. But this state of knowledge on nuclear interactions was not the end. Complex behavior is usually an indicator that the underlying basics are not yet revealed, like with the magnetic force that is in reality the electric force with relativistic effects. Further investigations showed that the real origin of the strong nuclear interaction is within the nucleons, and that what is observed between the nucleons is a residual effect; this is why it is today often called residual strong nuclear interaction. What had happened?

In 1968, it was discovered that the protons and neutrons that make up the nucleus consist of even smaller particles (Fig. 1-1), called **quarks**. And it was discovered that the fundamental interaction is actually between the quarks, while the interaction between the nucleons being composed of quarks is only a residual effect. This is somehow comparable to the interaction between electric dipoles, which is actually based on the interaction between the charges that make up the dipoles. The interaction between quarks is also associated with something like a charge, called color (Green 2006). A proton or a neutron consists of three quarks, and thus a proton or neutron has a multi-pole field. In addition, quarks have mass and electric charge, and the electromagnetic interaction can be relevant and thus another cause for complex behavior of the interaction and forces between the nucleons. The interaction between the nucleons, which is then not a fundamental one, is now called **residual strong nuclear interaction**. The term **strong nuclear interaction** or just **strong interaction** is now used for the fundamental interaction between the quarks. What about the weak nuclear interaction? It is seen today as connected with the electromagnetic interaction; it is sometimes just called **weak interaction**, and together they are called **electroweak interaction**. Both interactions are still quite different from the other fundamental interactions previously discussed; unlike gravitation and electric interaction, their associated forces have a limited range and have no $1/r^2$ dependence. The level of complexity is beyond the scope of this book; for more, Walker et al. 2014, wikipedia 2017b, wikipedia 2016a, and especially learner.org 2017 might be useful sources.

Besides for its complexity, **nuclear energy** will not be discussed any further here for another reason. Up to this point in section 2.2, all fundamental interactions were associated directly with a corresponding force and corresponding energy, and thus discussed individually. This is also possible for the nuclear interactions when looking at the fundamental ones. But the fundamental nuclear interactions do not show up alone; they are e.g. accompanied by another significant fundamental interaction, the electromagnetic interaction. Thus, for practical reasons, the term nuclear energy is commonly used different, combining different effects as nuclear binding energy or as nuclear reaction energy. This will be discussed briefly again in sections 3.1.6, and finally in detail in section 3.2.5. Thus, there is no detailed treatment here of the energy associated with the nuclear interactions individually, and this is also why nuclear energy is labeled in the heading with "…".

2.2.5 Energy conservation and conversion – more examples

Section 2.1 discussed energy conservation and conversion as based on few simple principles, mainly Newton's 2nd and 3rd law, focusing on simple situations of 1 object in motion, and cases of interaction with a 2nd object. The objects were considered point-like, and the interaction was just gravitation, thus the treated energies were kinetic energy and gravitational energy. In section 2.2 then, objects with an extension were included, and electromagnetic and nuclear interactions were discussed in addition. Altogether, the discussion thus covered all fundamental energy contributions.

The following examples, still using simple situations of 1 object in motion and cases of interaction with a 2nd object, show how these increase the variety of situations that can be described by the concept of energy. However, the strength of the concept of energy is not in describing simple situations where writing down the equations using forces is still possible, convenient, and giving insight. The strength of the concept of energy is in describing situations that are complex; so complex that their detailed description is inconvenient, often impossible e.g. if it is not understood in sufficient detail. But before going to complex situations in chapter 3, it is helpful to study first a few more simple situations by forces and energy in parallel.

A way to find more examples for energy conservation and conversion, which are easy to understand, is modifying the examples discussed before.

2.2.5.1 Objects that are point-like

Kinetic and gravitational energy – non-linear motion

Let's first start with examples that are still within the scope of section 2.1, thus point-like objects having kinetic and gravitational energy. But let's now look at non-linear motion that we avoided earlier to keep things simple.

We have discussed orbits e.g. of the earth around the sun, for convenience as being circular. Then E_{kin} as well as E_g are constant individually. What if this is not the case? We can start looking at this by an interesting situation that people had in mind when developing a description for gravitation: the flight of a bullet fired by a gun. As Fig. 2-35 (left) shows, and everybody knows, a bullet fired will fly for some time and eventually hit the ground.

Compared to the free fall (Fig. 2-11) there is now also the motion of the bullet due to being fired by the gun. How far it gets depends on its initial velocity and the angle between gun and earth surface. Of course there is the friction of the air slowing down the bullet, but let's neglect it. What happens? The force of gravitation pulls the bullet towards the earth, but if fired at the right velocity the bullet is pulled to the earth in a way that the overall result is a circular orbit. And if fired at a higher velocity? Then it becomes elliptic. And if the velocity is still higher, then the bullet even leaves the earth orbit.

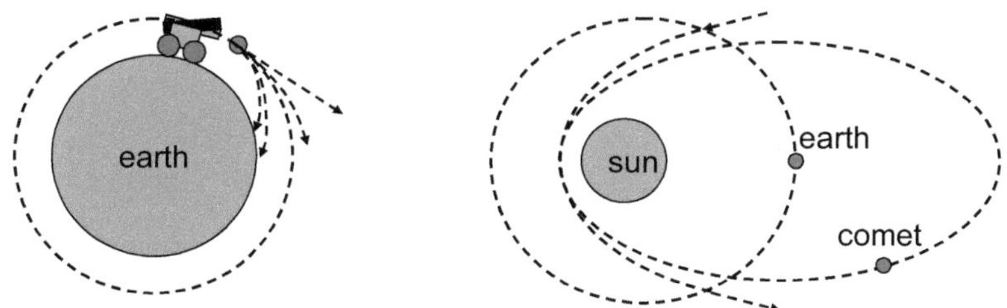

Fig. 2-35 Gun firing a bullet at various velocities (left) and elliptic motion of the earth and a comet around the sun (right)

The orbit of the planets around the sun is actually also not exactly circular, as discussed before, but instead rather elliptic. Kepler was the first who discovered and described the elliptic motion by his laws. What about energy? In an elliptic orbit the distance varies; with it varies the gravitational energy, and due to energy conservation also the kinetic energy. When a planet in an elliptic orbit gets closer to the sun it "falls" towards the sun and thereby gains kinetic energy at the expense of gravitational energy. An elliptic orbit is a permanent conversion between kinetic and gravitational energy in the system of the two objects. That the earth moves around the sun for millons of years is an example from common experience showing that friction can be neglected in space (vacuum) and that energy is conserved.

The next example is a bit similar, and extends the view on collisions between two objects. The term collision is commonly used for a special type of interaction, one being extremely short-range, repulsive, without further details. We discussed elastic collisions, meaning collisions without losses, before (Fig. 2-11). If we consider again the above orbits, an object entering the solar system (Fig. 2-35, right) could approach the sun, make a turn, and leave the solar system again e.g. in the opposite direction somewhat like in a collision, but now with a specified, long-range attractive interaction.

When a spacecraft passes a moving planet, it can however not only change its direction, but also increase its velocity. The procedure, called swing-by, employs the gravitational pull as Fig. 2-36 shows. Overall energy is conserved, but it is more useful to see the planet as doing work on the smaller spacecraft, thereby transferring part of its kinetic energy to the spacecraft.

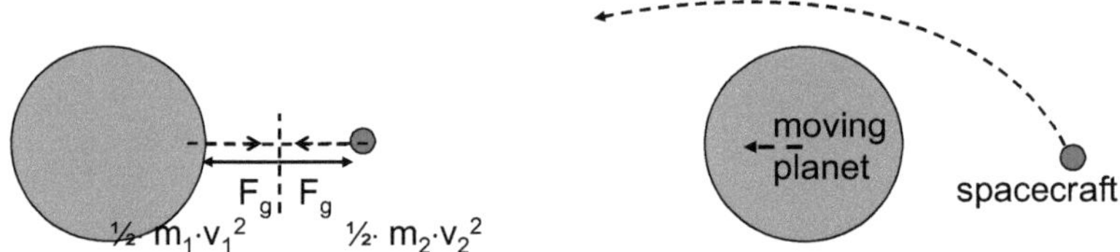

Fig. 2-36 Massive, moving planet accelerating a passing satellite

In a next step, we can now replace the gravitational interaction by the electric interaction, allowing then in general attractive and repulsive interaction.

Kinetic and electric energy – without and with conversion

Fig. 2-37 shows examples that were discussed in section 2.1.4.3, where a force had no effect on the energy such that energy was conserved, however now the force of gravitation is replaced by an electric force.

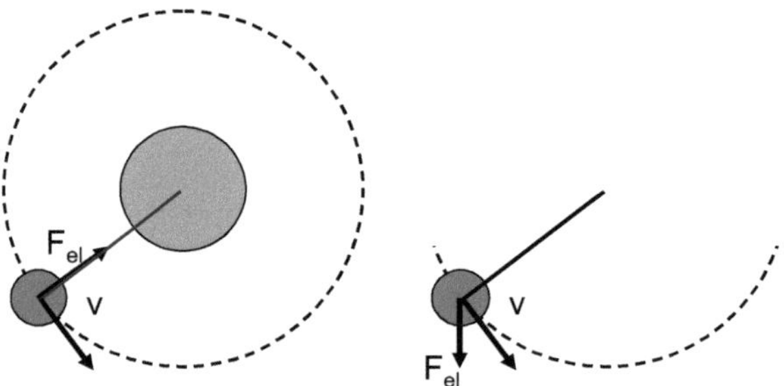

Fig. 2-37 Orbit and pendulum with electric instead of gravitational force

The example on the left shows two electric charges of opposite sign that attract each other such that one moves around the other one in a circular orbit. This was previously the earth in its orbit around the sun where the force of gravitation between sun and earth was the reason for the orbit. Strictly speaking, as charges are always in connection with mass, in a real situation gravitation would also be present. However, it is not necessarily relevant, e.g. in the orbit of an electron around the nucleus in an atom (Fig. 1-1). The example on the right shows what was previously the pendulum with the mass exposed to a gravitational force and bound by a rope; now it is a charge exposed to an electric force that is due to another charge not shown. While the total energy is conserved, conversion between electric energy and kinetic energy and back takes place repeatedly. The force of the rope has no energetic effect. Again, as charges are always in connection with mass, in a real situation this would also include gravitational energy.

The following examples are based on examples discussed in section 2.1.3.2.

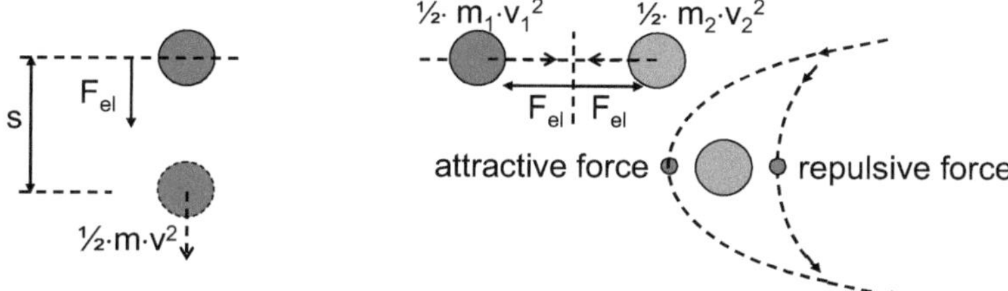

Fig. 2-38 Free motion of a charge in an electric field (left), and "collision" of two charged particles (right), possibly leading to open orbits

The free motion of a charge in an electric field shows Fig. 2-38, left. Now the free motion, which was with gravitation the free "fall", leads to a conversion between electric and kinetic energy, while their sum, the total energy, is conserved. Another way of looking at the situation is that the electric force does work on the charge and thereby changes its kinetic energy.

Shown in Fig. 2-38, right, is the "collision" of two particles, where the previously unspecified force is now a repulsive electric force. When the charges approach each other the repulsive electric force slows them down; thereby, their kinetic energy decreases while the electric energy increases. After the closest distance has been reached, the repulsive force will lead to an increase of the kinetic energy while the electric energy decreases with increasing distance. Another way of looking at the situation is that one of the particles via the electric force does work on the other particle. The collision is then seen as an exchange of energy between the particles via work.

Electric and gravitational energy without kinetic energy

Let's now discuss some examples that show the conversion between different potential energies, and without kinetic energy to make things easier. For this, we look at the quasi-static displacement between equal forces, where no net force exists and kinetic energy can be eliminated from the discussion.

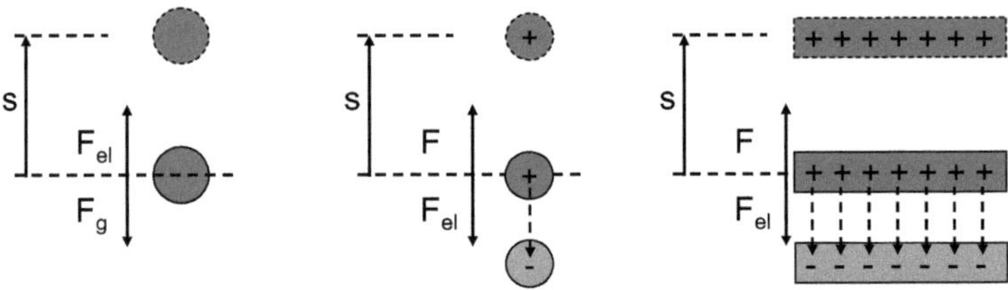

Fig. 2-39 From left to right: a charged object subject to gravitation and an electric force is relocated a distance s, the same with an unspecified force and an electric force, and finally the same for many charges

Fig. 2-39, left, shows a charged object subject to gravitation and an electric force of equal magnitude and opposite direction, thus that $F_{el}+F_g = 0$. When relocated by a distance s, the relocation is connected with the conversion between electric and gravitational energy while overall energy is conserved according $F_{el}{\cdot}s+F_g{\cdot}s = 0$. This can also be seen as the electric force doing work on the charge, thereby increasing its gravitational energy.

Fig. 2-39, center, shows a charged object subject to an electric and an unspecified force of equal magnitude and opposite direction, thus that $F_{el}+F = 0$, being relocated by a distance s. The relocation is still subject to energy conservation $F_{el}{\cdot}s+F{\cdot}s = 0$, but with an unspecified force it is more common to see it as the unspecified force doing work on the charge, thereby increasing its electric energy (with respect to the second charge of opposite sign that is shown explicitly in this case).

Fig. 2-39, right, shows the same situation for an object with many charges; but this is a simple case where the charges can also be viewed as a single, larger charge, and without taking into account the extension. This is why it was placed here as an example that shows the transition to more complex examples. The more general discussion for this follows in section 3.1.3.

It is possible to include gravitation in the examples at the center and right; then, work is done increasing the sum of electric and gravitational energy.

2.2.5.2 Objects with extension

Examples for the conservation and conversion of energy involving dipoles can also be constructed from the situations discussed before. For homogeneous fields there is no torque for mass and gravitation (Fig. 2-24), however there is one for the electric (Fig. 2-27) and for the magnetic case (Fig. 2-33).

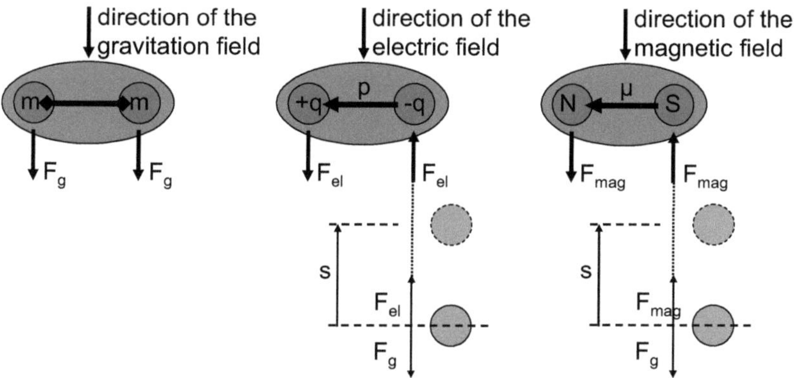

Fig. 2-40 For homogeneous fields there is no torque for mass and gravitation, while there is one for the electric and magnetic case. As shown on the bottom, a force being part of the torque can be used to lift an object against the force of gravitation.

As Fig. 2-40 shows, a force being part of the torque can be used to lift an object against the force of gravitation; energy is conserved, while electric energy or respectively magnetic energy is converted to gravitational energy. But both cases can also be seen as the dipoles doing electric work, respectively magnetic work, to lift the mass and thereby increase the gravitational energy of the mass. And of course dipoles can also interact with other dipoles and thereby affect them and exchange energy (Fig. 2-34).

This completes the examples to exercise again what was discussed about basics in section 2.1, and to explain how the basic energy contributions just introduced in section 2.2 extend the variety of situations covered by energy.

It must be stressed again that all these situations can, and often are discussed without talking about energy. Knowledge of the forces and Newton's laws is sufficient. This will change in the following chapter, chapter 3, when discussing "Complex situations with many objects".

2.2.6 Summary and conclusions

Let's now summarize the main points from chapter 2, as a repetition, and also to have a quick reference for the further discussion.

Chapter 2 discussed simple situations, which is 1 object in motion, and cases of interaction with a 2nd object. Section 2.1 introduced basic terms, concepts and laws. The objects were considered point-like, and the interaction was just gravitation. Because of these simplifications, the basic terms, concepts and laws can be understood (section 2.1.7). What we are interested in, in simple situations, is the description of the motion and position of objects, or changes thereof (and corresponding for complex situations, as we will see in chapter 3). Throughout history, the concepts of "force", "momentum", and "energy" were developed to describe the effect of an interaction between objects on their motion and position. The most direct concept, closest to common experience, is that of forces. The only origin of forces are interactions; forces just describe the effect of an interaction on objects. Using the concept of forces we can just look at a single object: Newton's 2nd law connects the force acting on an object with the change of its motion. Newton's 3rd law states that the force acting on an object is always due an interaction with a 2nd object, which is subject to a force of the same magnitude but opposite direction. And if there is no interaction, nothing changes, as expressed by Newton's 1st law. While it is the changes that we can see, it is the interaction that is the cause. Thus, the core of it all are interactions. An interaction is always between 2 objects, and comprises an action and a reaction, as is common experience, and expressed by Newton's 3rd law. The interaction between objects, affecting the objects (relative) motion and position, is the basis and thus common core in the concepts of "force", "momentum", and "energy". The balance between action and reaction shows up in all, and rewritten, results in the conservation laws. The interaction, and its effect on motion upon a change in position, can be expressed in any concept, in different ways. But the concepts are not fully equivalent. Forces give the most detailed information, and forces also relate directly to our common experience. For this, energy changes by a force acting a distance have a special name: work. Then, why use the concept of energy? The advantages of the concept of energy will become clear when dealing with complex situations in chapter 3. The basis is what we learned already, from simple situations; so let's summarize the crucial points about energy.

The concept of "energy", like of "force" and "momentum", describes the effect of an interaction between objects on these objects motion and position. At the core of the concept of energy are, as in other concepts, interactions. An interaction always acts on both objects that interact with each other, and the balance between action and reaction is showing up in all the concepts. Energy conservation is understood by looking at single interactions: action and reaction result in a pair of energy changes that have equal magnitude and opposite sign such that together energy changes cancel out. Thus, it is better to say that energy does not change, to point to the physical origin. Saying that energy as a value is conserved is correct, but it hides the physical origin, and leads to the belief that energy is a "thing" that is conserved. Energy conversion is due to a "connection" between different energy forms (appearance of energy). For example, an objects mass connects its kinetic energy by its motion with its gravitational energy by its relative position.

In section 2.2, objects could also have an extension, not just point-like, and the remaining fundamental interactions, namely electromagnetic, strong and weak nuclear interaction were discussed (Fig. 2-41). Besides the fundamental interactions, we discussed two more types of interactions where the forces are unspecified: by "shelf", "hand" or any similar, and by collision. But these are still due to one or a combination of fundamental interactions. A **collision** of two objects is a special type of interaction, in a dynamic situation, by a repulsive but otherwise unspecified force, extremely short-range to act at "direct contact". In an **elastic collision**, the kinetic energy of the colliding objects is conserved, meaning no "losses" occur by conversion to another energy form. In an **inelastic collision** losses occur.

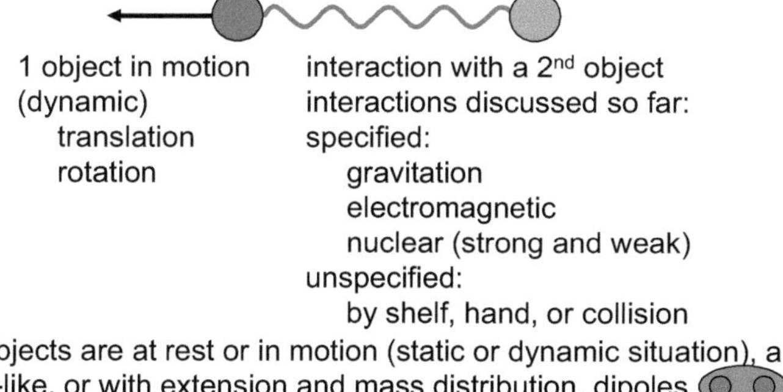

Fig. 2-41 General situation in simple situations with 1 or 2 objects

From the fundamental interactions, the next step is the fundamental energy contributions. The "fundamental energy contributions", as discussed in section 2.2, refer to simple situations, which are 1 object in motion, and cases of interaction with a 2nd object, and comprise potential energy due to one of the 4 fundamental interactions, and in addition of course kinetic energy.

Tab. 2-1 gives a full overview including when objects are point-like or have extension. Consequently, besides location and translation, orientation and rotation can then play a role. For the interactions the simple cases of dipoles in a homogeneous field are given. There is no relation to the size of objects; the objects can be small, called **microscopic**, to large, called **macroscopic**. The nuclear interaction is however not relevant for large objects because the interaction is limited to distances within atomic nuclei.

Tab. 2-1 Fundamental energy contributions for simple situations, which is 1 object in motion and cases of interaction with a 2nd object; details

Energy of motion of an object, called **kinetic energy**, connected to a	
change of	**location** (translational motion) $E_{kin,\,trans} = \frac{1}{2} \cdot m \cdot v^2$
	orientation (rotational motion) $E_{kin,\,rot} = \frac{1}{2} \cdot \Theta \cdot \omega^2$

Energy of interaction with a 2nd object, called **potential energy**, connected to the	
relative	**location**, originating from the
	gravitational force, simplified $E_g = m \cdot g \cdot s$
	electric force, simplified $E_{el} = q \cdot E \cdot s$
	strong and weak nuclear force
	orientation, originating from the
	electric force on an el. dipole $E_{el} = -p \cdot E \cdot \cos\varphi$
	magnetic force on a mag. dipole $E_{mag} = -\mu \cdot B \cdot \cos\varphi$

The 4 fundamental interactions take a key role in energy conversion. All objects subject to a fundamental interaction have mass, and thereby connect kinetic energy and gravitational energy. An electric charge in addition connects electric and magnetic energy. They connect the changes of distance or orientation, and explain why the concept of energy relates so many different changes, while momentum does not.

Regarding key idea number 3 from section 1.1 "The origin of energy conservation and conversion is in simple situations, which is 1 object in motion, and cases of interaction with a 2nd object", the discussion is now complete.

The discussion also showed some common misconceptions about energy:

The term potential energy is often used synonymously with gravitational energy, but there are other potential energies, and they include also effects associated with orientation (angular position). But looking at daily life the limited statement is at least not wrong; nobody sees nuclear energy, magnetic energy besides the compass, or electric energy besides the sticking of fluff to clothes.

Energy due to any of the 4 fundamental interactions is not a "property" of an object! All contributions of potential energy are due to interactions with another object and thus depend on the relative position; for gravitational energy we need to keep in mind that it is due to the interaction with the earth, and not just the gravitational energy of the object as is commonly said. But what about kinetic energy? This leads to another crucial point. To say that the interaction energies are potential energy, meaning "hidden", just refers to the visual observation. However, regarding analysis of the concept of energy it is rather the opposite: it is the interaction energies where the basis of conversion and of conservation becomes clear, and thus the basis of the whole concept, while kinetic energy is not giving any insight. And kinetic energy also depends on the chosen reference frame; it becomes a value by selecting a reference frame e.g. in which it interacts with another object. From a scientific point of view, the "origin" of energy is in the interactions and changes of relative position between objects, while kinetic energy is better seen as energy that is intermittently "stored" between interactions.

Further on, the common statement that there is no absolute energy is obviously wrong. It originates from energy conservation and conversion, which state that changes in energy in different energy contributions / forms are interrelated and cancel out; this in fact does not imply an absolute energy. However, kinetic energy is zero at zero velocity, and the potential energies are zero when the force becomes zero at infinite distance. This is not obvious from energy conservation, or the work - energy theorem (and later the 1st law of thermodynamics). However, they do not say anything about it; they only deal with changes, and this has probably led to the misconception.

Let's now, after discussing the individual fundamental energy contributions, look at the whole concept of energy. First, Tab. 2-2 shows again a summary of the fundamental energy contributions, shortened and grouped differently: electric and magnetic energy are separate, according common experience. In section 3.2.5 we will see another difference for the nuclear energy.

Tab. 2-2 Fundamental energy contributions for simple situations, which is 1 object in motion and cases of interaction with a 2nd object; short

energy contribution / form				
motion	interaction			
kinetic	gravitational	electric	magnetic	nuclear

Because the fundamental energy contributions cover everything, and do not count anything twice, they are a complete set. Therefore, the total energy is the sum of the energy in all fundamental energy contributions

$$E_{tot} = E_{kin} + E_{pot}$$
$$= E_{kin} + E_g + E_{el} + E_{mag} + E_{nuc} ,$$

Eq. 2-57

and energy conversion is between them. The related energy changes are

$$dE_{tot} = 0 = dE_{kin} + dE_g + dE_{el} + dE_{mag} + dE_{nuc}$$
$$= d\left(\frac{1}{2} \cdot m \cdot v^2\right) + d\left(\frac{1}{2} \cdot \Theta \cdot \omega^2\right) + m \cdot g \cdot ds + \dots .$$

Eq. 2-58

Eq. 2-58 shows clearly what was already concluded before: the concept of "energy", like of "force" and "momentum", describes the effect of an interaction between objects on these objects motion and position. Energy correlates the changes of velocity v, angular velocity ω, relative position s …

And work? In simple situations, which is 1 object in motion, and cases of interaction with a 2nd object, the energy of an object can be changed by exchange of energy by an interaction with another object (Fig. 2-41), in other words the surrounding. An "exchange" of energy by a force acting a distance is called work. The process is described by the work - energy theorem

$$W = \Delta E .$$

Eq. 2-59

Taking the origin of the work into account energy is again conserved.

The discussion up to now was limited to simple situations, which is 1 object in motion, and cases of interaction with a 2nd object by one of the fundamental interactions. These limitations were necessary to stay with situations that are easy to comprehend, and showed that the concept of energy is not needed in simple situations; neither Newton nor Kepler used it. In simple situations the equations to describe the situation can be written down using forces, and using forces things are also easy to comprehend. In chapter 3, the limitations are removed and complex situations with more objects, and stepwise also more than one fundamental energy contribution are discussed.

Actually, we have already looked at some slightly complex situations. In Fig. 2-10 we looked at a pendulum interacting with another by a collision. Each pendulum is actually a situation with two objects, as the earth must be taken into account as partner in gravitation. Then, as shown in the overview in Tab. 2-3, pendulum 1 having kinetic and gravitational energy exchanges energy with pendulum 2 having also kinetic and gravitational energy by an interaction in form of a collision. Another example was lifting an object against the force of gravitation, shown in Fig. 2-11 on the right. Again, the situation already comprises the object and the earth, and is affected from the surrounding by an interaction with whatever lifts the object. And similarly, the same holds for the situations with dipoles, shown in Fig. 2-40.

Tab. 2-3 Overview on options for slightly complex situations

energy changed		interaction	energy changed	
E, abs.	E_{kin}	collision ...	**E, abs.**	E_{kin}
	E_g	g		E_g
	E_{el} (pos. + orient.)	el.		E_{el} (pos. + orient.)
	E_{mag} (orient.)	mag.		E_{mag} (orient.)
	E_{nuc}	nuc		E_{nuc}

This already indicates that the concept of energy relates an incredible variety of effects related to changes in the motion (linear and angular) and position (location and orientation) of objects due to interactions. This is what makes energy as a concept so useful, but on the other hand also so hard to understand when many objects and several interactions are involved.

3 Complex situations with many objects

Chapter 2 discussed energy related to simple situations, which is 1 object in motion, and cases of interaction with a 2nd object by one of the 4 fundamental interactions. For these situations, energy conversion and conservation are easy to understand. Kinetic energy and gravitational energy of macroscopic objects relate to common and everyday experience such that we are familiar with them and understand them, and based on them, electric and magnetic effects look familiar, and to some degree even nuclear effects. For simple situations, dealing just with 1 or 2 objects, it is possible and actually common to describe things without talking about energy, its conversion and conservation. This allows to understand why energy is conserved and how energy is converted. Thus, simple situations were chosen as introduction.

The term energy came to common use when the situations were complex, and the origin of the effects not understood, e.g. thermal energy. Then, conservation and conversion of energy, and what energy is itself, is "mystical". The strength of the concept of energy is not in describing simple situations; here writing down the equations using forces is possible, convenient, and moreover giving insight. The strength of the concept of energy is in describing complex situations; so complex that their detailed description is inconvenient, often even impossible because the situation is not understood in sufficient detail, or because it cannot be calculated with sufficient accuracy. But if we do not understand what is going on in detail in complex situations, how can we understand why energy is conserved, and how it is converted? The answer is actually easy. We know that energy is conserved in a single interaction and how it is converted. So we know it also for two interactions that happen in series, or in parallel. And thus also for many interactions.

Chapter 3 will now show that energy conservation and conversion in complex situations with many objects and even more than one fundamental interaction can still be explained and understood, and that conservation can be proved as long as Newton's laws are applicable. Beyond the validity of Newton's laws the derivation does not hold, such that the derivation by Newton's laws is not a general proof. Nevertheless, in many cases it can be argued why energy conservation still holds.

To discuss complex situations with many objects we replace the "1 object" from the discussion of simple situations in chapter 2 (Fig. 2-41) now by "many objects" (Fig. 3-1). To simplify things, "many objects" are often called a "system". A **system**, in general the sum of parts that interact with each other, is a number of objects, a quantity of matter, or a region of space. The **boundary** of a system separates the system from its **surrounding**, and can be a real or just an imaginary surface. There are three types of systems. An **open system**, also called a **control volume**, is a system that allows mass and / or energy exchange across its boundary. A **closed system**, also called a **control mass**, is a system with no mass exchange across its boundary. And an **isolated system** is a system that has no mass and in addition also no energy exchange across its boundary, thus its mass and energy are constant.

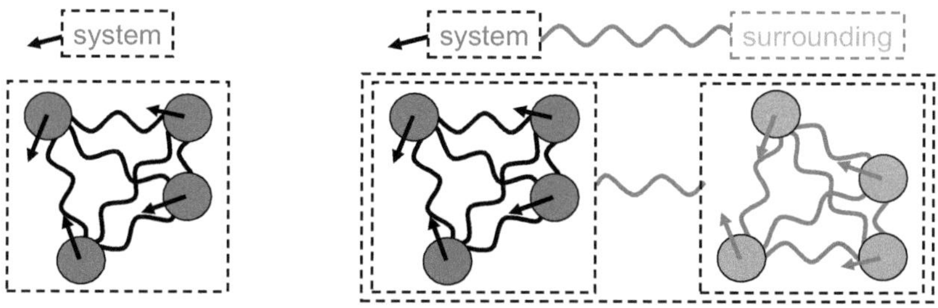

the objects are at rest or in motion (static or dynamic situation), and point-like, or with extension and mass distribution, dipoles ...

Fig. 3-1 General situations with many objects: a system of objects without (left) and with (right) an effect from the surrounding

The discussion of complex situations is split into two, general situations. The first general situation (Fig. 3-1, left) is a system of many objects in motion and / or interacting with each other with no effect of the surrounding; then, energy is conserved. The second general situation (Fig. 3-1, right) is with an effect, by an interaction with the object(s) of the surrounding; again, overall energy is conserved, but individually, the system and its surrounding can exchange energy by their mutual interaction. Thus, the two general situations follow the discussion of the work - energy theorem, but compared to the discussion in chapter 2 (Fig. 2-41) now for many objects in motion and / or interacting with each other. The effect from the surrounding can be discussed e.g. as interaction with the object(s) of the surrounding by a force (field), by impulse, or in the concept of energy as work and heat.

The system of "many objects" can have energy by its overall motion, or its overall interaction with the object(s) in its surrounding; this is like before. The system of "many objects" has then energy of the basic energy contributions by its overall motion or position of the system as a whole. However, compared to the discussion in chapter 2 (Fig. 2-41), replacing "1 object" by a system of "many objects" (Fig. 3-1) adds a new, very important aspect. The system of "many objects" can possibly also undergo internal changes; this was not the case with "1 object" before. Energy corresponding to the internal motion or mutual interaction of the objects is called **internal energy**. Changes of the internal energy can be due to an effect from the surrounding if the effect causes internal changes. Energy corresponding to the overall motion or interaction could be called **external energy** (Mehling 2016).

Section 3.1 continues the discussion of the fundamental energy contributions from section 2.2, now for many objects; it discusses the cases of kinetic energy and of potential energy due to a single fundamental interaction, for point-like objects as well as for objects with an extension. At the end of section 3.1, for any number of objects all fundamental energy contributions are defined and discussed, so no part is missing or counted more than once. This is a crucial necessity to use the concept of energy, especially its conservation, for calculations, predictions, etc. If the set of energy contributions would be incomplete, or parts would be counted more than once, then an energy balance calculation could lead to erroneous results.

But there is a serious problem, and this is dealt with in sections 3.2 and 3.3: the fundamental energy contributions are not all connected to what we can observe, and many things we observe are not described by them directly. For example, we cannot observe changes in the energy due to the nuclear interactions directly, as the affected particles and distances are far too small. And chemical and thermal energy, which are directly observable by a change in composition or by a change from solid to liquid, are not only associated with effects on the atomic and molecular level that we do not see, the observed changes are also by several fundamental energy contributions. We observe macroscopic changes, realize that they are related to energy, and can even measure that energy is conserved, but we cannot see the microscopic origins that explain why it is conserved and how it is converted. This is what made the concept of energy so "mystical", but at the same time so useful: it connects changes in a large variety of things and allows predictions of changes even without understanding why and how things work.

This multitude of things, hard to understand in detail, is why energy is such a complex but also helpful and successful concept.

But with the discovery of the microscopic effects, starting with the kinetic theory of gases, explanations for many observations started to become available. They are used in section 3.2 and 3.3 to discuss effects related directly to observations and to explain them. As already discussed in section 2.1.4, here the difference between an energy contribution and an energy form becomes clear. An energy contribution can be anything, for example the kinetic energy of the electrons in an atom, while an energy form refers to contributions that have a significant meaning in science, technology, or everyday life, related to a directly observed change, e.g. thermal energy.

And chapter 3 will even explain more crucial things. Like Newton's 1st law is not what we directly observe due to friction, as Galilei already realized, it is the same with energy conservation. Energy sometimes seems to disappear, or come out of nothing. For example, when a ball rolls and due to friction finally comes to rest, or when a ball falls to the ground and comes to rest; where is the energy? It must be somewhere! There must be energy forms we can't see, or more precise, we do not notice the related changes. Friction we tried to avoid up to this point in the discussion. In the discussion of Newton's laws it causes a force. In macroscopic observations it seems to be a non-conservative force, but it must be related to an energy contribution. So, when kinetic energy is "lost" by friction, where does it go?

A central point is also the discussion of the 1st law of thermodynamics, formulated by R. Clausius (1882 to 1888) as "the change of the internal energy of a system is equal to the amount of heat supplied to the system, and less the work done by the system". Therefore, the work - energy theorem is not only extended to many objects and also to include more energy contributions / forms; thermal energy is one of them, and as a consequence the work side is extended by heat. Work and heat are no energy contributions / forms; work and heat are instead energy exchanged between a system and its surrounding, which is in fact itself another system (Fig. 3-1).

These are the things discussed in the following sections, discussing complex situations with many objects, including several fundamental interactions. Thus, chapter 3 is the crucial step in understanding what is commonly often not understood about energy. And it is the basis to the treatment in science and technology today, as discussed in chapters 4 and 5.

3.1 Energy related to a single fundamental energy contribution

Section 3.1 discusses cases of a single fundamental energy contribution for systems of many objects, without or with an effect by interaction with object(s) of the surrounding. Strictly, all objects can interact with each other; for simplification, Fig. 3-2 does not show all mutual interactions between objects of the system with object(s) of the surrounding. The effect from the surrounding is discussed as interaction, or by the corresponding force field.

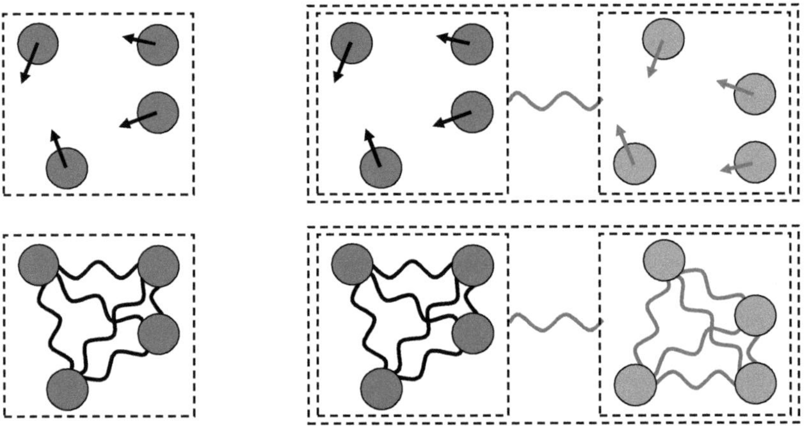

the objects are at rest or in motion (static or dynamic situation), and point-like, or with extension and mass distribution, dipoles ⬭ ...

Fig. 3-2 General situations with many objects: a system of objects without (left) and with (right) an effect from the surrounding; the top shows the case for motion only, the bottom for interaction only

3.1.1 Kinetic energy

Kinetic energy is associated with all objects having a mass being in motion, and only with them (thus, gravitation is also present and could be relevant). Kinetic energy is the least complex and thus the easiest to understand; and it is experienced in everyday life. Put simply, the discussion of kinetic energy in complex situations with many objects is the same as for simple situations with only 1 or 2 objects, just with more objects. So, it is the perfect start.

3.1.1.1 Situations without / with an effect from the surrounding

Let's start looking at a situation without an effect from the surrounding to allow focusing just on the system of many objects. The kinetic energy of a system of many objects can refer to objects of any size, point-like or with extension and mass distribution, from small particles to stars and galaxies.

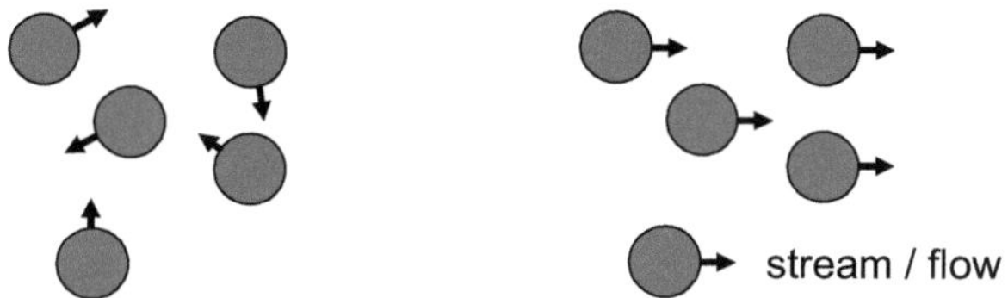

Fig. 3-3 Motion of many objects by a linear translation with arbitrary (left) or the same (right) direction

Fig. 3-3 shows the motion of many objects by a translation with a velocity of arbitrary (left) or same (right) direction, possibly of different magnitude. An example of a system with arbitrary direction is an ideal gas, where the particles (objects) are point-like (diameter negligible to the average distance), with no relevant forces between particles except elastic collisions; thus, there is no potential energy. If all objects move in one direction, it is called a flow or stream, e.g. a flow of sand (grains), or air (molecules).

Fig. 3-4 Motion of many objects by a rotation of the individual objects (left) or all objects as a whole (right)

Fig. 3-4 shows the motion of many objects by a rotation of the individual objects (left) or all objects as a whole (right). Again, direction and magnitude can be different. The rotational energy of the individual objects is only possible / relevant if they have significant extension and mass distribution. For example, dancers on a dance floor can rotate pairwise, and air (molecules) in a tornado or hurricane rotate around a common center.

Let's now look at the situation with an effect from the surrounding. For a system of objects having kinetic energy, this can be an interaction with an object having also kinetic energy, by a collision with one or several of the objects of the system. Examples of such situations are in pool billiard, bowling, and many other similar games, often with the objects of the system initially at rest. If the object from the surrounding looses kinetic energy, the system gains kinetic energy, and if it gains kinetic energy then the system looses kinetic energy. Overall energy is conserved.

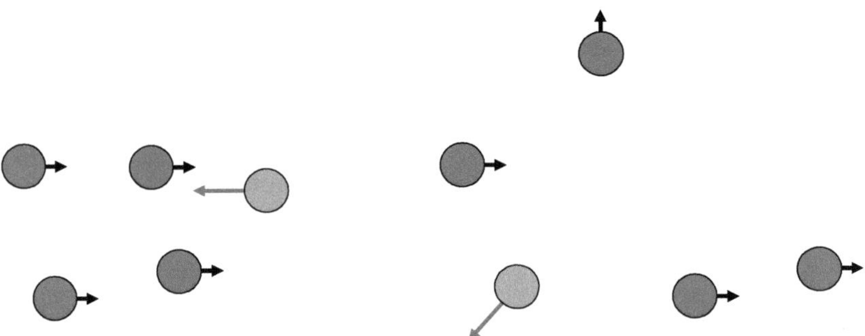

Fig. 3-5 Interaction of an external object by a collision with one or several of the objects of the system before (left), and after collision (right)

Translation and rotation are independent energy contributions to kinetic energy, and it is possible that several contributions come together. Thus, the **kinetic energy** E_{kin} of a system of objects is the sum of all its contributions, taking into account all objects (marked i) and translational as well as rotational contributions

$$E_{kin} = \sum_i E_{kin,i} = \sum_i E_{kin,trans,i} + \sum_i E_{kin,rot,i}$$
$$= E_{kin,trans} + E_{kin,rot} \; .$$

Eq. 3-1

It is common to call kinetic energy an energy contribution or an energy form depending on the situation. If we can see the motion, like in the case of a moving car, it is an energy form. For electrons moving in a metal, where we cannot see the motion, it is an energy contribution and in this case to the energy form of thermal energy. The difference will become clearer with more examples. In any case, translational kinetic energy and rotational kinetic energy are energy contributions to kinetic energy.

3.1.1.2 External energy without / with an effect from the surrounding

If an object moves free, like a stone thrown in the air, it has a point that moves smoothly, called its center of mass CM, while its rotation is around that point; e.g. for a ball the CM is at the center of the ball (section 2.2.1). The same applies for a system of objects. The system's kinetic energy is the sum of the kinetic energy associated with the motion of its center of mass, thus the system as a whole, and the kinetic energy associated with the motion of the individual objects of the system relative to the center of mass, thus motion "in" the system.

The first, the kinetic energy associated with the motion of its center of mass, thus the system as a whole, is its **external energy** (here of kinetic energy). The system as a whole has a mass and a rotational inertia, and its external energy is thus given by the translational motion of its center of mass, and its rotational motion around it. This was already discussed in section 2.2.1.1 and section 2.2.1.2. The relevant terms are thus all from the case without an effect from the surrounding. Neglecting internal changes, the system is like a single, rigid object. External kinetic energy is for example the kinetic energy associated with a mass of air (particles) moving in a direction or with a moving box, of a body of flowing water, or for a single object as just a ball. Its overall kinetic energy can change by a collision of the whole system, e.g. when a ball bounces back from the ground the kinetic energy of the ball including the air molecules inside can change.

3.1.1.3 Internal energy without / with an effect from the surrounding

The kinetic energy associated with the motion of the individual objects of the system relative to the center of mass, thus motion "in" the system, is its **internal energy** (here of kinetic energy). The system's objects each have a mass and a rotational inertia. Its internal energy is given by the translational motion of its objects with respect to the center of mass, and respectively for the rotational motion. Again, this was already discussed in section 2.2.1.1 and section 2.2.1.2 for the individual objects. The relevant terms are thus all from the case without an effect from the surrounding. Internal kinetic energy would be for example the kinetic energy of the gas particles enclosed in a box, which is at rest.

3.1.1.4 Energy conservation and the work - energy theorem

If there is no effect from the surrounding the internal energy is conserved. Why? In simple words, this has been said before. We know that energy is conserved in a single interaction, e.g. when two objects collide (elastic collision), as long as Newton's laws are valid; then, the acting and reacting force in an interaction lead to an action and reaction energy that cancel out, as was proved in section 2.1.3.2. Under these conditions we know energy is also conserved in a situation with many objects, no matter if the many individual interactions are in series or parallel. This way of explanation is much easier than making a calculation of the development of the motion of the many objects with time and calculating the associated energy. This would also show that the energy is conserved, but it is a lot of work. The motion after scattering is hard to calculate, and for example for a gas with billions of atoms or molecules a calculation is impossible. Anyway, why should we? We know the result already. And this way of explanation is also more informative than observing energy conservation by an experiment and measurement of the energy. In short, as there is no effect from the surrounding, the situation here is an isolated system; it has no energy exchange at its boundary, and inside its energy is conserved.

If there is an effect from the surrounding the energy of the system can change, e.g. by a collision with an object from the surrounding. When taking the surrounding into the energy balance however again energy is conserved; this is because the system together with its surrounding is again an isolated system and the arguments from above apply.

The exchange of energy of the system with its surrounding can now again be described by the work - energy theorem. At the beginning of this chapter, it was already said that the second general situation (Fig. 3-1, right) under investigation, the one with an effect from the surrounding due to interaction with one or many other objects, relates again to the work - energy theorem. However, compared to the discussion in chapter 2 (Fig. 2-41), it is now for many objects in motion and / or interacting with each other.

In section 2.1.3, energy conservation in simple situations was discussed and then the work - kinetic energy theorem was introduced. It states that the work W done on the object under investigation by a force F acting a distance s results in the change of its kinetic energy. Both are energies, and equal due to overall energy conservation. Work just denotes the energy exchanged between the system and its surrounding (e.g. another system) by a force acting for a distance, and the change in kinetic energy is the resulting effect on the system. This is simple to understand for a single object, and is a common experience with large objects. Later the theorem was extended to include besides kinetic energy also potential energy (Eq. 2-23), becoming the work - energy theorem.

Now, what about a system with many objects? Without effect from the surrounding, all forces are internal from collisions between the objects of the system, and from Newton's 3rd law they are pairwise and cancel out. Thus, there is no net force and consequently the energy within the system is conserved (replacing ds by dt and dE_{kin} by dp the same holds for the linear momentum). So, in an isolated system there is no energy exchange across the boundary, meaning no work is done on the system, thus that the systems internal as well as external kinetic energy is constant. As an example, take a box filled with an ideal gas. The number of particles is so large that we cannot calculate things in detail. But we know that if the box is initially at rest, it will stay at rest. The kinetic energy of the system as a whole does not change by particle collisions, so the box cannot start to move by collisions of the particles with the walls. And its internal energy also stays the same.

But what if there is an effect from the surrounding? An interesting case shows Fig. 3-6: a system of two objects, which are initially at rest, collides with a moving object from the surrounding. All objects have the same mass, and the collision (elastic) is just with one object such that the object from the surrounding comes to a halt while the object of the system that took part in the collision takes up the kinetic energy. As a result, the center of mass of the system moves such that it has gained external energy as kinetic energy, and at the same time, the objects of the system move away from the center of mass such that there is also now internal energy as kinetic energy.

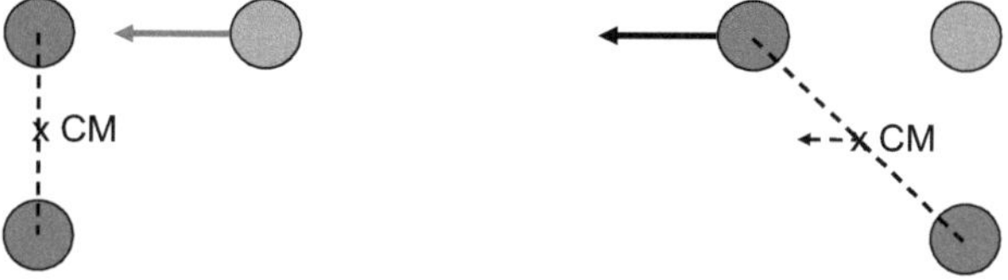

Fig. 3-6 A system of two objects of equal mass, initially at rest, collides with an object from the surrounding

This process is easy to understand regarding the work - energy theorem. Neglecting the second object of the system that is not involved in the collision, this is a collision of two objects as already discussed in section 2.1.3. The object from the surrounding does work on the system and as a result its kinetic energy decreases while the kinetic energy of the system increases. Taking into account the second object of the system that is not involved in the collision does not make a difference. The system of objects interacts with its surrounding resulting in an exchange of energy. Is it work? Work was defined as a force acting a distance, and the collision in Fig. 3-6 can be described that way. This even holds if the object from the surrounding does not come to a halt but is instead redirected. The work done on the system by the object from the surrounding and the work done by the system on the object in the surrounding together are zero due to energy conservation.

What if more than one object of the system is involved? This is the case in pool billiard where all objects have the same size. Fig. 3-7 shows a similar case: a larger object interacts with a system of small objects initially at rest.

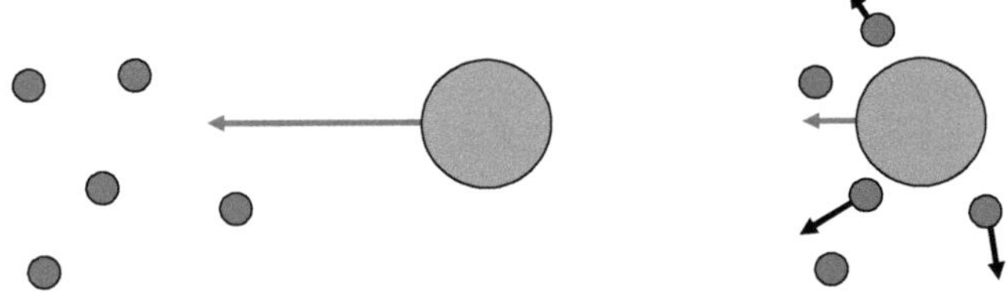

Fig. 3-7 Larger object interacting with a system of smaller ones; initial situation and after few collisions

The collisions lead to a loss of velocity and thus energy of the larger object, while the small objects gain velocity and thus energy. If all collisions are subject to Newton's laws, energy conservation is still valid and proved by the derivation discussed for the simple situation. This shows once more the crucial advantage of the concept of energy, especially energy conservation and the description of an exchange of energy by the work - energy theorem. Just by knowing how much energy is exchanged between the system and its surrounding it is also known (from energy conservation) how much the energy of the system changes; there is no need to calculate the final velocities of all objects from the initial velocities and energy exchange by collision. But it does not give the absolute value of the energy (nor does it need its initial value). Knowing the energy exchanged tells the systems change of energy (internal and / or external), but not its absolute value. Is the exchange of energy still as work? An important point is that the directed motion of the large object is reduced, while the undirected motion of the many small objects increases. Think about pool billiard. There is a significant difference if the system contains more than 1 object, and if the collisions with the external object involve several of them: while a single collision is reversible, multiple collisions become less and less reversible. The undirected motion in pool billiard will never make the first ball move back. This cannot be expressed by an exchange of work, meaning a force acting a distance. For the moment, we will delay this however to section 3.3.3, where it will lead us to another form of energy: thermal energy.

But this is not the end of the story about kinetic energy. If the smaller objects are invisible, then kinetic energy of the large object, easily visible, ends up as not visible motion of the smaller objects, and thus seems "lost". Macroscopically, this effect is called **friction**; energy is still conserved, but the direct observation of the large object tells a different story. The force exerted by the small objects on the larger object, called friction force F_f, makes the larger object slow down. The associated energy "loss" is $F_f \cdot ds$ where ds is the distance the friction force acts. In classical mechanics, the force of friction is called a **non-conservative force**, as it seems to violate energy conservation; classical mechanics does not include the energy form that describes where the "lost" energy has gone. But it is possible to get an idea of what is going on; when an object moves through water it causes convection, meaning water currents that circulate and thus rotational kinetic energy. With time they become smaller and smaller until they disappear. Where do they go? They become kinetic energy of the particles of the water, the water molecules, which are invisible. And if an object moves through air it is the same; we just never see the air currents unless we use smoke or anything similar to make them visible. The energy ends up in another energy form: thermal energy, which we discuss in section 3.3.3.

It is an incredible achievement of Galilei to conclude correctly that gravity works the same on all objects, and that feathers fall slow due to air resistance even though he could not see what goes on in detail. His conclusion that it is just friction that makes the ball stop rolling, and that without friction it would roll forever is the basis for Newton's 1st law. It is nothing less than the one fundamental basis for a correct concept of forces and energy.

3.1.2 Gravitational energy

Gravitation acts between all masses, it is always attractive, and it depends only on the relative location (distance) of the objects, not on their motion. This makes it the second least complicated case, and the next to discuss. In situations with many objects all objects interact with each other, and as the force of gravitation is always attractive this results in an overall attraction.

3.1.2.1 Situations without / with an effect from the surrounding

Gravitation is common in everyday life, and situations of a system without / with an effect from the surrounding are familiar to everybody.

It is gravitation that allows us to stand on the surface of the earth, as well as buildings etc. And even the earth, the planet we stand on, wouldn't be there without gravitation. The most important example for a static situation, thus without kinetic energy, with many objects interacting by gravitation is the existence of the planet we live on! The earth was formed from smaller objects due to their mutual attraction by gravitation (Fig. 3-8), and this is still what keeps the planet together, including also the oceans and atmosphere.

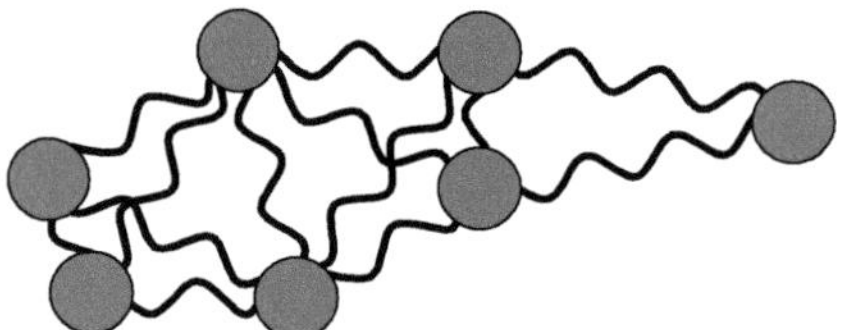 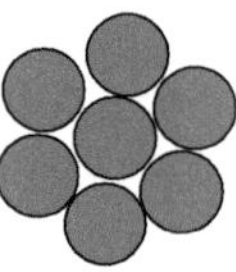

Fig. 3-8 Formation of the earth from smaller objects due to their mutual attraction by gravitation (left) and final result (right)

And when you lift a stone from the ground you take a piece of the planet, pull it away from the rest, and you feel the rest of the planet pulling it back by the gravitational force (Eq. 2-37) of its parts; it is the combined effect of the mass of the rest of the planet. When you hold just two stones next to each other, you do not feel any force between them; it is just too weak. Strictly speaking, you or the stone act also upon the earth and can be viewed as an object of the surrounding of the earth, causing an effect on the earth. However, there are more obvious examples. We see them in the sky, like Galilei when he discovered the moons of Jupiter.

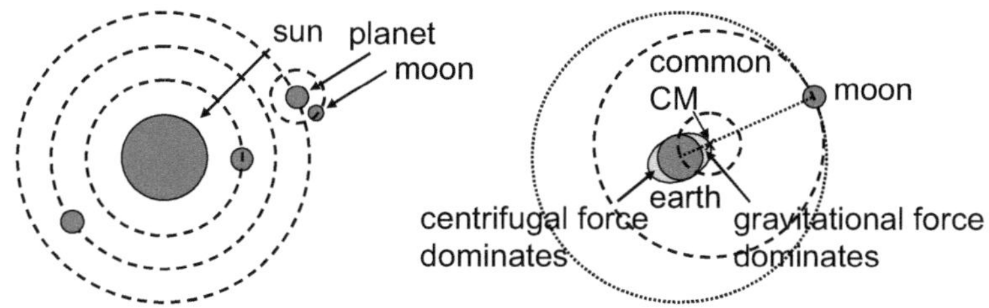

Fig. 3-9 Left: planets, moons, and the sun in the solar system; right: earth-moon system and causes for the tides

Because mass is not only the basis for gravitational but also kinetic energy, both often come together. Examples for dynamic situations are the planets, moons, and the sun in the solar system as a whole (Fig. 3-9, left), where gravitational energy and kinetic energy of translation and rotation is present. Looking at the system of a planet and its moon, the sun is an object in their surrounding, causing an effect, here the orbit, by the force of gravitation.

3.1.2.2 External energy without / with an effect from the surrounding

What are examples for external energy in the case of gravitational energy? A system of objects as a whole has a mass, which is the sum of the masses of its objects, like the mass of the earth is the sum of the masses of its parts. When neglecting internal changes, the system is like a single rigid object. As discussed in section 2.2.2, as an interaction energy, gravitational energy exists only in combination with another object with mass interacting with the system by gravitation. This holds of course also if a system comprises many objects. This is the big difference between kinetic energy and the interaction energies; kinetic energy as external energy exists by the motion of the system itself, without any effect from its surrounding.

The value of the external **gravitational energy** is given by the location of the system in the gravitation field caused by the other object in its surrounding (Eq. 2-41); there is no energy related to its orientation if the gravitation field is homogeneous in the region of the system (section 2.2.2.4). There is no need to discuss the case of an inhomogeneous field here; however, it can easily be understood by just looking at all masses and forces individually.

3.1.2.3 Internal energy without / with an effect from the surrounding

And internal energy? In general, the internal **gravitational energy** E_g of a system of objects is the sum of the gravitational energies of all objects (marked i, j) by their mass due to their mutual interaction

$$E_g = \frac{1}{2} \cdot \sum_i \left(\sum_j E_{g,i,j} \right) \quad i \neq j \,.$$

Eq. 3-2

Here $E_{g,i,j}$ is E_g from Eq. 2-41 for the masses m_i and m_j at a distance of $r_{i,j}$; the summation has to be done without counting interactions twice (thus ½) and without self-interactions (thus $i \neq j$). For example, the internal gravitational energy E_g of the system "earth" is the sum of the gravitational energies of all its parts.

If the system "earth" is subject to a gravitational force due to another mass in its surrounding, e.g. the moon, then the earth as a system has external energy with respect to the moon. But the internal energy of the earth is not affected by the moon, as long as the gravitation field by the moon is considered homogeneous (at a close look this is not exactly the case, which is causing the tides). In the same way, the system of planet and moon shown in Fig. 3-9 has gravitational energy E_{int}, and it has also gravitational energy E_{ext} with respect to the sun.

3.1.2.4 Energy conservation and the work - energy theorem

There is no need to discuss here in detail why energy is overall conserved; but it is clear from Eq. 2-41 and Eq. 3-2 that a direct calculation is difficult. If kinetic energy is also involved the situation is even worse. Already a situation with 3 objects having an interaction energy like gravitation and at the same time kinetic energy has no general analytic solution. An example of such a problem, also called **3-body problem**, is the sun, earth, and moon. The development of the motion and position of the objects with time can only be calculated numerically, such that energy conservation cannot be shown directly from a detailed calculation. However, knowing that energy conservation holds for a single interaction between two of them, it must also hold in general; we thus do not need to calculate anything at all.

102

What if a system is subject to gravitation by an object in its surrounding? The object in the surrounding exerts a directed force of gravitation, and when the distance of the system changes, by work done on or by the system, then its external gravitational energy changes, as discussed in section 2.2.2. And what about its internal energy? If the force field of gravitation caused by the object in the surrounding is homogeneous, meaning the same throughout the system, then its effect is the same on all parts such that no internal changes occur. For example, the orbit of the moon around the planet in Fig. 3-9 is the same without or with sun. But if the field is not homogeneous? If looking at things in detail, this is the case for the system of the earth with its oceans which are affected by the moon (Fig. 3-9, right). The earth and moon move around their common CM, and the result of the small difference on opposite sides of the earth of the gravitational force of the moon and the **centrifugal force** F_c (section 3.2.1.1) results in the tides. The gravitational force of the moon not only makes the water in the oceans move as a whole, it even affects the "solid" earth.

3.1.3 Electric energy

As already discussed in section 2.2.3, the electromagnetic interaction can be separated into two cases: an electric interaction when the charges are at rest, and when the charges are in motion an additional, magnetic interaction. Both were discovered separately, as the observed effects are very different. And because we try to describe observations, they are treated separately. This also considerably simplifies formulas, and is also common practice in science and technology. Let's start with the electric interaction.

The electric interaction and its related force acts between all charges, and it can be attractive or repulsive. All charges also have a mass and are thus also subject to gravitation in real situations. The electric interaction is dominating in situations involving objects with a significant net charge and / or small mass. This is the reason why we rarely see its effect directly on macroscopic objects. First, let us discuss the electric interaction and associated force and energy. The electric force and energy depend on the location of the charged objects, and when talking about dipoles on their orientation. They do not depend on motion; this is included in magnetic energy.

Now, neglecting the gravitational interaction due to the mass of the charges, what happens in situations with many charged objects?

3.1.3.1 Situations without / with an effect from the surrounding

Fig. 3-10 shows a situation with many objects, single charges (or dipoles), interacting by the electric interaction without an effect from the surrounding. All objects interact with each other, like in the case of gravitation; but in contrast, the electric interaction can be attractive as well as repulsive. This situation is the basis for the inter-particle interaction in materials, specifically between atoms and molecules that have become ions (Fig. 3-13, left), or electric dipoles (Fig. 3-14). Including motion, it is also the basis for the interaction in metals (Fig. 3-13, right), and between the pos. charged nuclei and the neg. charged electrons in atoms (Fig. 3-28).

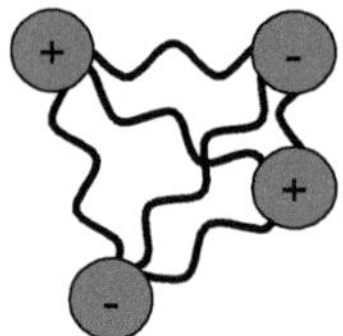

Fig. 3-10 Situation with many objects, single charges, interacting by the electric interaction without an effect from the surrounding

Very important are situations with many objects, single charges or dipoles, interacting by the electric interaction with an effect from the surrounding. Fig. 3-11 shows only a single charge in the surrounding, and a single mark to represent its interactions with all objects / el. charges of the system.

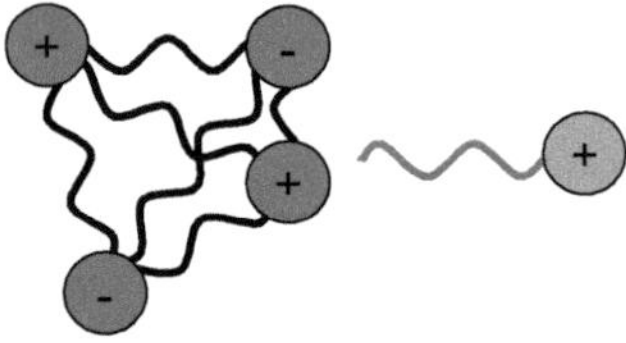

Fig. 3-11 Situation with many objects, single charges, interacting by the electric interaction with an effect from the surrounding

This situation is the basis for electric currents, and thus electric applications. It is also the basis for the interaction of electromagnetic waves with matter (section 3.1.5), specifically light. The electric interaction is commonly described by its electric force or force field, or its energy field (el. potential).

3.1.3.2 External energy without / with an effect from the surrounding

Now, what is the external electric energy of a system of charged objects? The system of electric charges as a whole can have a net charge, but also a net dipole moment. Neglecting internal changes, the system is then like a single rigid object with a single net charge or single net dipole moment. This case was already discussed in section 2.2.3 as an interaction energy, electric energy as external energy exists only in the presence of an electric field, e.g. caused by another object with a charge or a dipole moment and thereby interacting with the system.

The value of the external **electric energy** depends on the position of the system in the electric field as discussed before, specifically for system with a net charge and subject to a homogeneous electric field in section 2.2.3.3, and if the system has a net dipole moment and is subject to a homogeneous electric field in section 2.2.3.4.

3.1.3.3 Internal energy without / with an effect from the surrounding

And what about the internal electric energy of a system of charged objects? Using the similarity to gravitation, the internal **electric energy** E_{el} of a system of charged objects is the sum of the electric energies of all charged objects (marked i, j) in the system due to their mutual interaction

$$E_{el} = \frac{1}{2} \cdot \sum_i \left(\sum_j E_{el,i,j} \right) \quad i \neq j .$$

Eq. 3-3

Here $E_{el,i,j}$ is E_{el} from Eq. 2-46 for the charges q_i and q_j at a distance of $r_{i,j}$; the summation has to be done without counting interactions twice (thus ½) and without self-interactions (thus $i \neq j$).

A detailed calculation is obviously involving quite some effort, even if the number of charges in the system is small. However, it is possible to make some general conclusions that give important insights without calculations. According Eq. 2-46, in general a high internal energy results from charges of equal sign being close to each other, and from charges of opposite sign being far away. Accordingly, a low internal energy results from the reverse.

Let's look at two important cases: individual charges, and pairs of charges with equal magnitude and opposite sign (thus zero net charge), which are called dipoles. Fig. 3-12 shows some situations with overall zero net charge.

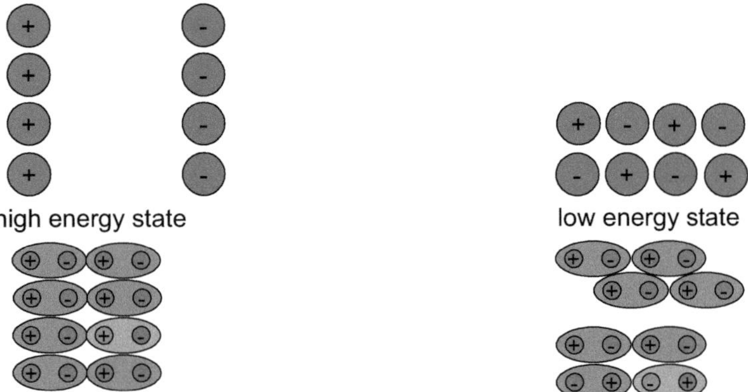

Fig. 3-12 Examples of situations with high and low energy states for individual charges (top) and dipoles (bottom)

For individual charges (Fig. 3-12, top), a state with high energy is when all positive charges are on one side and all negative charges on the other some distance away; then, charges of equal sign are close to each other, while charges of opposite sign are far away from each other. A low energy state is for example when a negative and a positive charge build a pair of charges that is then neutral. When allowed free to move, the charges will try to reduce the internal energy of the system to a lower state.

Where is this relevant? Large objects usually have no significant net charge, unless artificially created, so for large objects the electric force rarely plays a role compared to other forces. To find examples we look at small scales. An example for a static situation is the situation in salts (Fig. 3-13, left). In a **salt**, the atoms or molecules are electrically charged by loosing one or several electrons to neighbors or by gaining one or several electrons from them. The charged atoms are called **ion**; specifically, the positively charged ones are called **cation** and the negatively charged ones are called **anion**. The electric force between them is so strong that the ions stay together, usually in a regular pattern, which is called an **ionic crystal**. The interaction is called **ion - ion interaction**, and the related bonds are called **ionic bond**.

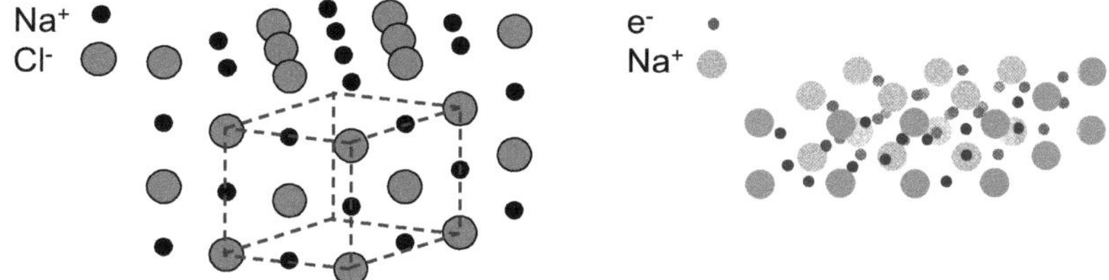

Fig. 3-13 NaCl as an example for a salt (left), and Na as an example for a metal (right)

Examples for dynamic situations, where besides electric also kinetic energy plays a role, are charges in an ionized gas, which is also called a **plasma**. And a somewhat mixed situation is in metals. In a **metal** (Fig. 3-13, right), when solid, the atoms have become ions by giving away one or more electrons, in contrast to a salt however not to a neighbor atom. Instead, the electrons occupy the space between the metal cations and are free to move, which is why they are also called **electron gas**. The electron gas is the basis for electric conduction. The corresponding bond is called **metallic bond**.

For dipoles (Fig. 3-12, bottom), the high and low energy states can be derived from the states for individual charges, taking into account that the two charges in a dipole are a fixed pair. It is important that at close distance the electric fields of dipoles are inhomogeneous, thus causing reorientation as well as attraction between them despite having a zero net electric charge.

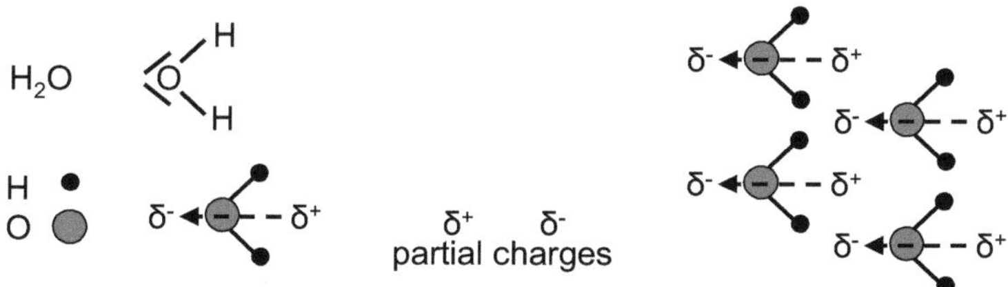

Fig. 3-14 Water molecule as dipole, and dipole-dipole interaction

Examples are e.g. substances with polar molecules, like water (Fig. 3-14). In a water molecule (H_2O), the oxygen atom attracts the bond electrons a bit more than the hydrogen atoms, such that a water molecule is a small dipole. In a system with many water molecules, the molecules stay together due to the interaction between their dipoles, called **dipole - dipole interaction**.

An important point is that in the high-energy states on the left in Fig. 3-12 there are many equal charges on the surface of the system; this means, the corresponding surface has a net charge and the system is like a large dipole; it is then said to be **polarized**. This is the case if the objects are dipoles, but also when they are just single electric charges.

How can the system, with individual charges or with dipoles, get to such a high-energy state? Of course only if it is forced to. The most straightforward option is by an electric force acting on the system. So, let's now look at the case where the system is affected by an electric field, e.g. caused by an electric charge or dipole in the surrounding. The electric field from the surrounding (Fig. 3-15) results in a directed force, in addition to the forces between the charges or dipoles in the system, which has an ordering effect on the negative and positive charges in the system, and the same for dipoles. Single charges will tend to move such that the positive ones move in direction of the electric field (which e.g. originates from a positive charge and ends at a negative), while negative charges move in the opposite direction. A change in the location of charges or orientation of dipoles in the system changes the internal state and thus also the internal energy of the system.

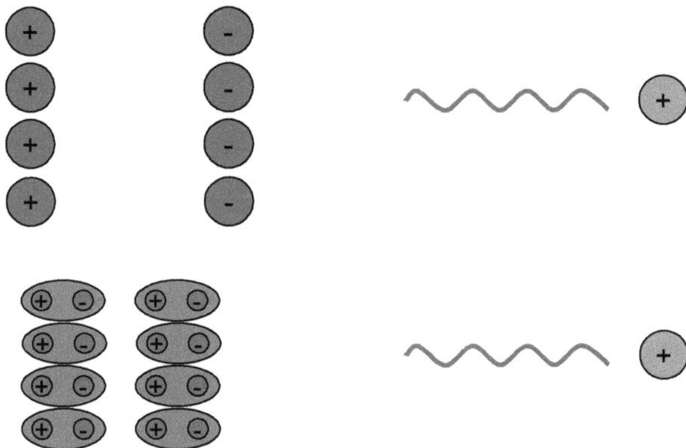

Fig. 3-15 Examples of situations with high energetic states

A comparison of Fig. 3-15 and Fig. 3-12 shows that with the electric field, due to its associated directed force acting on the charges in the system, the high-energy state from the situation without an effect from the surrounding is favored. The same way, dipoles tend to reorient in a way that a high-energy state is favored. As the orientation of a dipole is from its negative to its positive charge, the dipoles tend to get oriented with the external electric field, as was already observed for single dipoles in section 2.2.3.4. In both cases, individual charges or dipoles, the system is polarized by the electric field from the surrounding; thus, polarization does not need preexisting dipoles. For mass and gravitation such an effect doesn't exist, at least not in a homogeneous field; the reason is that the same force acts on all objects.

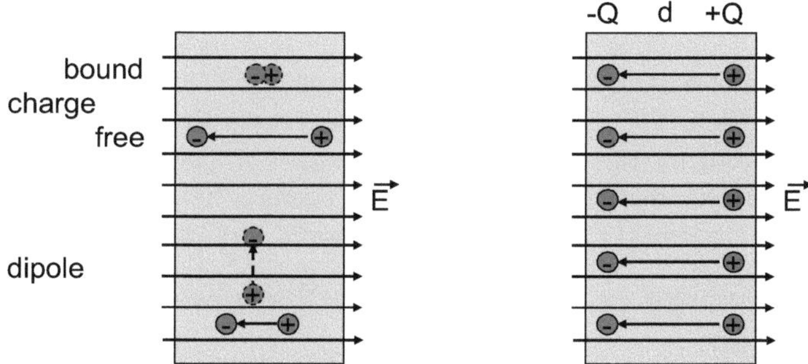

Fig. 3-16 Left, effect of a directed electric field on charges and dipoles in a material, and right, resulting polarization

Let's be more specific now and look at a **material** and the charges within it. The charges can for example be electrons and nuclei in atoms in general, or anions and cations in a salt, or molecules with a dipole moment like water. If the charges are bound, they can only change their average position until the force due to the external electric field E is equal to the opposing force of the bonds. Single charges simply move somewhat (Fig. 3-16, left top) while dipoles reorient (Fig. 3-16, left bottom). The net effect is that positive and negative charges, no matter if they are individual charges or part of a dipole, are somewhat moved in an ordered way; this is called the **dielectric effect**. The overall effect is the same as if a charge Q is present at the surface, called **surface charge**, separated by a distance d (Fig. 3-16, right) equal to the thickness. The product of both is called the **polarization P**

$$P = Q \cdot d .$$

Eq. 3-4

What is the internal energy in this state? It is, by Eq. 2-47,

$$E_{el} = P \cdot E .$$

Eq. 3-5

Using Eq. 2-47, the electric field E is the field inside that we don't know. To calculate E_{el} directly would require to know the position of all charges and dipoles, and to apply Eq. 2-46 and Eq. 3-3; this can be done for simple cases, but usually is impossible. We will see shortly however that it is easy to calculate the change of the internal energy, not its absolute value, using the work - energy theorem. We will later see that in most applications we do not need the absolute value anyway, just the change of the internal energy.

But before, let's complete this discussion by looking at another case: the charges are not bound, such that they can move freely, e.g. like the electrons in metals. Then, the external electric field causes the charges to move, maybe affected by effects similar to friction. The net motion of electric charges is called an **electric current**, and its control by the electric field applied to the material is the basis for all electronic applications.

3.1.3.4 Energy conservation and the work - energy theorem

There is no need to spend much time discussing if energy is conserved or not. The same holds for the electric interaction as before for gravitation: energy is conserved in a single interaction as long as Newton's laws are valid; then, the acting and reacting force in an interaction lead to an action and reaction energy that cancel out. Under these conditions, we know energy is also conserved in a situation with many objects. So, the energy in a system is conserved if there is no effect from the surrounding, and if there is one, and energy is exchanged between the system and its surrounding, then energy is conserved taking the system and its surrounding together. There is no need to look at the location of the charges and orientation of the dipoles in detail.

Let's now get back to the question from above, which was how to calculate the change of the internal energy of a system of charges or dipoles when an electric field is applied from the surrounding; this is where we will see the advantages of the work - energy theorem. Fig. 3-17, on the left, shows again a situation that we discussed already before in section 2.2.5.1 (Fig. 2-39): an unspecified force F separates positive charges from negative charges further by a distance ds, thereby doing work $dW = F \cdot ds$ on the system of charges and in turn increasing its internal energy by $dE_{int} = dE_{el} = F_{el} \cdot ds$. This is just applying energy conservation and the work - energy theorem, nothing else.

Instead of doing this by an unspecified force from the surrounding, we want to do this now by an electric force. The change of the internal energy is still $dE_{int} = F_{el} \cdot ds$, where F_{el} is the force acting in the system between the charges. The work done is however now not $dW = F \cdot ds$ with an unspecified force, but instead $dW = F_{el} \cdot ds$ with F_{el} in this case being the electric force applied from the outside. To avoid kinetic energy we assume again that the forces balance, and just look at a relocation. Then, from the work - energy theorem, and from F_{el} and ds being the same on both sides, follows that $dE_{int} = F_{el} \cdot ds$ where $F_{el} \cdot ds$ is from the surrounding. Thus, we can determine the change of the internal energy by the change done in the surrounding.

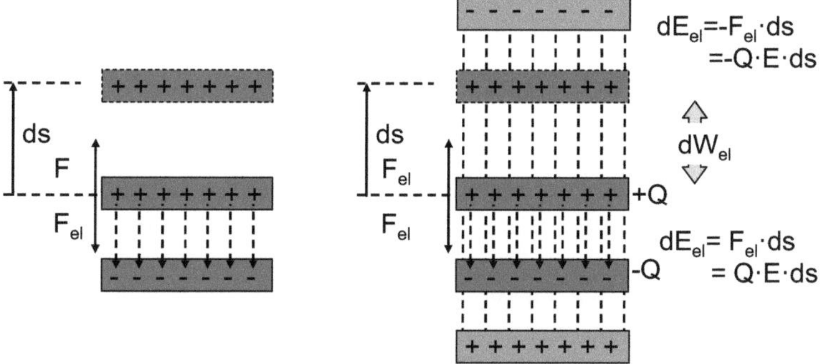

Fig. 3-17 Left: an arbitrary force F separates positive charges from negative charges by a distance ds, thereby doing work on the system and increasing its internal energy; right: the same by an electric field

Let's now include the electric field and charges into the calculation. The system we look at, shown on the right in Fig. 3-17, is still composed of two parts (darker shaded), one with positive charges and another with negative charges of equal magnitude. The cause of the force F_{el} from the outside to separate the charges is an electric field from the surrounding, which is due to another set of charges (brighter shaded). The charges are shown to understand the situation better. For convenience, the charges have the same magnitudes such that the fields and the forces have also equal magnitude; consequently, we do not need to consider kinetic energy. Now, when the part of the system with the positive charges is separated from the one with the negative charges further by a distance ds, the change of the internal energy of the system as calculated form its internal state is

$$dE_{int} = dE_{el} = \vec{F}_{el} \cdot d\vec{s} = Q \cdot \vec{E} \cdot d\vec{s} = \vec{E} \cdot d\vec{P} , \qquad \text{Eq. 3-6}$$

where E is the internal el. field, and the result is the same as by Eq. 3-5. This change is done by using energy from the surrounding, which is

$$dE_{sur} = dE_{el} = \vec{F}_{el} \cdot d\vec{s} = Q \cdot \vec{E} \cdot d\vec{s} = \vec{E} \cdot d\vec{P}$$
$$= dW_{el} . \qquad \text{Eq. 3-7}$$

and is equal to the work done by the surrounding on the system. Here, E is the external el. field. Eq. 3-6 and Eq. 3-7 look the same; however, in one case E denotes the electric field inside, in the other outside. Because in the example here both are equal, this is in agreement with energy conservation.

The case discussed here, where Q is constant and just moved by ds, is easy to understand; the energy change of the system can be calculated easily, and actually even its absolute value can be calculated. But what if in the system dipoles are reoriented? And what if we don't even know what is going on in the system? Already for Eq. 3-5 we noted that even if the polarization P or its change is known, we still need to know the internal electric field, which we usually do not know. The good news, as we saw, is that we do not need to know the details; this was already discussed in connection with Fig. 3-16. We only need to know the external electric field and the change of polarization. Using energy conservation, this allows to calculate the change of the electric energy in the surrounding by Eq. 3-7, which is equal to the **electric work** done on the system, and thereby also the change of its internal energy. But we only know the change of the internal energy, not its absolute value.

The discussion in section 2.2.3.4 has shown that Eq. 3-7 already includes the case of dipoles. This is also clear from the similarity of Eq. 3-7 and of Eq. 2-51. But why do they have opposite sign? For the single dipole discussed in section 2.2.3.4 as well as for the system of charges discussed here, the stable state is when the dipole(s) are oriented in the direction of the electric field; this reduces the energy outside, while increasing the energy inside the system. What is different isn't how the system reacts but our viewpoint. For the single dipole, described by Eq. 2-51, the focus is on the overall energy and not on the internal energy. This is the reason why it has an opposite sign "-" for changes of energy, and that this leads to neg. values in Eq. 2-51 is a freedom of normalization that can be used when looking only at changes of the energy.

Finally, what about work and energy if the charges can move free? Then, an electric field causes them to move. This charge transport, or charge flow, is called **electric current** I

$$I = dQ / dt,$$
Eq. 3-8

and simply the effect of a single object discussed before but now for many. We can then look at a fixed set of charges moving a distance ds, or a fixed distance where a number of charges move by. Both views are equivalent. From Eq. 3-7 follows

$$\frac{dE_{el}}{dt} = \frac{dQ}{dt} \cdot \vec{E} \cdot d\vec{s}.$$
Eq. 3-9

Using the **electric potential** U, with

$$U = \vec{E} \cdot d\vec{s},$$
Eq. 3-10

follows for the change of electric energy per time, called **electric power** P_{el}

$$P_{el} = \frac{dE_{el}}{dt} = \frac{d}{dt}(Q \cdot U) = \frac{d}{dt}(Q) \cdot U = I \cdot U.$$
Eq. 3-11

Summary and important conclusion of the preceding discussion

The discussion of the electric energy of a system of objects, which can be charges or dipoles, has shown something crucial, very different from before. Something that does strongly affect the discussion in the rest of the book.

Discussing the internal electric energy of systems with many objects, we saw examples that were so complex that the calculation of the absolute value is difficult, or even impossible. The reasons are too many objects, they are too small to be observed, or we do not have detailed information on their motion and interaction. Consequently, the same holds for a calculation of changes of internal energy if the changes are to be calculated as the difference of the internal electric energy before and after a change. But there is a solution. We saw that, employing the work - energy theorem, the change of internal electric energy can be calculated from the energy exchanged with the surrounding as work in an easy way, by the applied electric field and the change in polarization P. Thus, for convenience or lack of detailed information, we often just look at changes of energies, not absolute values.

3.1.4 Magnetic energy

Now, after finishing the discussion of the electric interaction, let's discuss the second part of the electromagnetic interaction: the magnetic interaction. Compared to the electric interaction, there is no magnetic charge, meaning monopoles; but there are magnetic dipoles (for simplicity, we do not discuss their origin in all cases, like spin, and also not multipoles like quadrupoles).

Neglecting the interaction by gravitation between the mass of magnetic dipoles, what happens in situations with many magnetic dipoles? In a system with many magnetic dipoles, the dipoles interact with each other and, if present, with a magnetic field from the surrounding. Actually, magnetic dipoles behave quite similar to electric dipoles.

3.1.4.1 Situations without / with an effect from the surrounding

Fig. 3-18 shows a sketch of a situation with many objects, now magnetic dipoles. Because there is no magnetic charge, there is no simple attractive or repulsive force like between masses or between electric charges. Instead, there is a torque between the magnetic dipoles, and if the magnetic field experienced by the individual dipoles is inhomogeneous, an additional attractive or repulsive force results. Both are used in many technical applications.

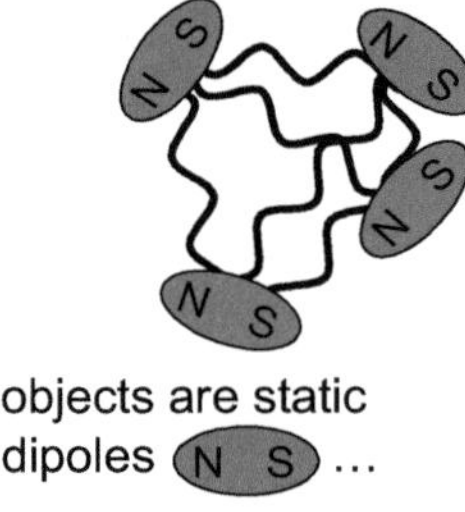

Fig. 3-18 Situation with many objects, magnetic dipoles, interacting by the magnetic interaction without an effect from the surrounding

The interaction of magnetic dipoles can be observed macroscopically on permanent magnets, and all of us usually have some experience with them. By analogy, the behavior of electric dipoles can also be understood easier. The general situation is not only the basis for the inter-particle interaction between macroscopic dipoles like compass needles. The magnetic interaction is also important and common on the microscopic scale where we don't see it: between the particles in atoms and molecules, as well as between the magnetic dipoles of atoms and molecules in materials.

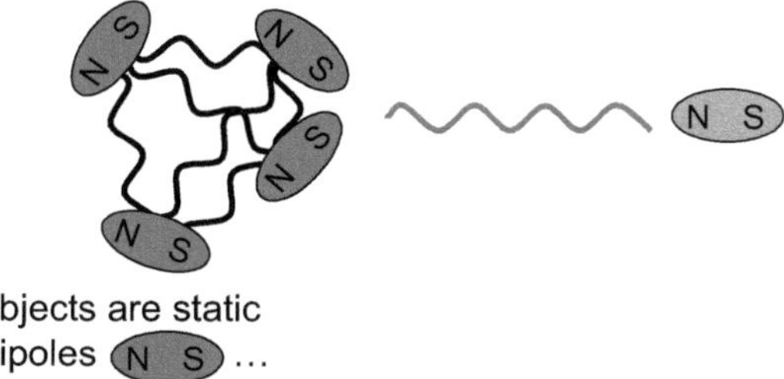

Fig. 3-19 Situation with many objects, magnetic dipoles, interacting by the magnetic interaction with an effect from the surrounding

The situation of a system of many magnetic dipoles with an effect from the surrounding by a magnetic field, e.g. due to another dipole, shows Fig. 3-19. It is technically important, as it describes the magnetization of materials by an electromagnet, thus increasing the magnetic effect of the electromagnet.

3.1.4.2 External energy without / with an effect from the surrounding

What is external energy in the case of magnetic energy? As there are no magnetic monopoles, the system we look at consists of magnetic dipoles; other multipoles we want to ignore for simplicity. And we also want to use this restriction now for the system as a whole. Neglecting internal changes, the system is like a rigid object with a single net magnetic dipole moment. External energy in the case of the magnetic energy only exists when the system is exposed to a magnetic field from the surrounding. This is the same as for gravitation and for the electric interaction.

The simple case of a magnetic dipole in a homogeneous magnetic field was already discussed in section 2.2.3; the field can be caused e.g. by another object with a magnetic dipole moment interacting with the system. For a homogeneous magnetic field, the external **magnetic energy** of the system is thus given by its orientation in the magnetic field. If the magnetic field is not homogeneous, the location of the system also plays a role; this is the basis for the attractive force between magnets.

3.1.4.3 Internal energy without / with an effect from the surrounding

In situations with many magnetic dipoles they all interact with each other. The **magnetic energy** E_{mag} of a system of magnetic dipoles is the sum of the energies of all magnetic dipoles (marked i, j) due to their mutual interaction

$$E_{mag} = \frac{1}{2} \cdot \sum_i \left(\sum_j E_{mag,i,j} \right) \quad i \neq j. \qquad \textbf{Eq. 3-12}$$

The summation has to be done without counting interactions twice (thus ½) and without self-interactions (thus $i \neq j$). The calculation of the individual interaction energies $E_{mag,i,j}$ is quite complicated; the previous discussion in section 2.2.3.8 only holds if the local magnetic field is homogeneous, which is only the case when the dipoles are at large distances. At close distance, the magnetic field of a dipole is inhomogeneous, as Fig. 2-32 shows. Thus, while the calculation of the magnetic energy is basically possible, even for the simple case of the interaction between two magnetic dipoles it is hard.

But we can already learn a lot from discussing some special cases. For example, Fig. 3-20 shows different situations with many magnetic dipoles and with effects of their mutual magnetic interaction on orientation and location.

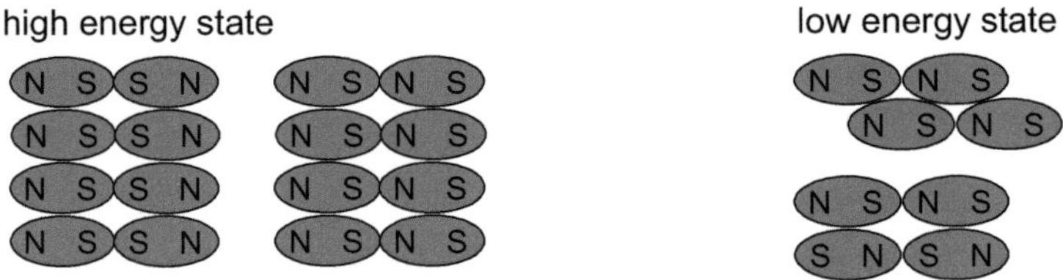

Fig. 3-20 Examples of situations with high and low energetic states

The high and low energetic states can be derived by looking at the states for electric dipoles (Fig. 3-12, bottom). In general, a state with high energy is when the interaction and its force is repulsive, thus when similar poles are close (N-N or S-S), while a state of low energy is when the interaction and its force is attractive, thus when opposing poles are close (N-S). A low energy state with N and S poles close, e.g. as shown in Fig. 3-20 on the right, is reached when the magnetic dipoles are allowed to move completely free.

A state with high energy where similar poles are close shows Fig. 3-20 on the very left: the case of two rows of dipoles, where in each row the dipoles are parallel such that equal poles are close, and also the poles of both rows being close are equal. Another, similar case is shown next to it: the case when the poles of both rows being close are opposite, such that the energy of the system is not as high. Therefore, some interactions are attractive and thus at a low energy, and some others are repulsive and thus at high energy. Assuming dipoles in one row are fixed by some effect, this orientation of dipoles in the second row is exactly what would be expected if they are considered individual and in the field generated by the others (section 2.2.3.8). In this case, there are many equal poles on the surface of the system; thus, the system as a whole has a net magnetic "polarization"; it becomes itself a magnetic dipole. The effect is the same as for electric dipoles, however, for the magnetic interaction a different set of terms is used; the system is said to be **magnetized** (instead of polarized), and has a net **magnetization** (instead of polarization). For experiments, an array with compass needles is useful. It visualizes the same way as Fig. 3-20 (center) that a large magnet, being a dipole, is composed of smaller dipoles oriented in parallel such that the magnetic effect (field and force) adds up. When a large magnet is split into parts, the parts are themselves magnets, just smaller; but they never become monopoles (this was already discussed in section 2.2.3.7).

Now, let's discuss the case with an effect from the surrounding by a magnetic field. This shows Fig. 3-21, with a single dipole as source of the magnetic field in the surrounding to indicate the direction of the magnetic field.

Fig. 3-21 Example of a situation with high energetic state

The situation is again similar to the electric interaction (Fig. 3-15), but now there is no case for charges, just dipoles. When exposed to a magnetic field from the surrounding, the result is an additional force on the dipoles in the system; it changes the internal state and thus internal energy of the system. For simplicity we assume the magnetic field is homogeneous; then the force is directed and has an ordering effect on the magnetic dipoles in the system.

Because the orientation of a dipole is defined from its S to its N (Fig. 2-32), the dipoles tend to become oriented with the external magnetic field as was already observed for a single dipole in section 2.2.3.8. If the external magnetic field is strong enough, it overcomes the local magnetic fields of the single dipoles and the dipoles become oriented in parallel (aligned), as Fig. 3-21 shows. In other words, the external field makes the magnetic field experienced by a single dipole in the system look almost homogeneous. From Fig. 3-20 we know that all dipoles aligned is a high-energy state of internal energy in the absence of a magnetic field from the surrounding. Thus, the magnetic field from the surrounding raises the internal energy of the system. The situation can also be observed nicely with an array of compass needles.

Now, let's be specific and look at materials. What are the magnetic dipoles? The electrons orbiting in an atom around the nucleus (Fig. 1-1) can be seen as an electric current loop, and as we previously discussed, an electric current loop is a magnetic dipole associated with a dipole moment (Fig. 2-32). There is also a second effect, the **spin**, e.g. of electrons; it is not possible to explain the spin by classical physics, thus its origin is not discussed here. For most atoms and ions, the magnetic effects of the electrons due to their spin and their orbital motion altogether exactly cancel out; such atoms have no intrinsic magnetic dipole moment. However, in atoms where the effects do not cancel out there is an intrinsic magnetic dipole moment, and different effects can be observed. If a net magnetic dipole moment is observed in the presence of an external magnetic field and disappears again when the field is removed, then the effect is called **paramagnetism**. If the effect remains at least partly after the field is removed it is called **ferromagnetism**. As a result, the observed state can differ even for the same external conditions; this is generally called **hysteresis**. After the external magnetic field has oriented the dipoles, in ferromagnetism they remain oriented due to a strong mutual interaction between the dipoles. Thus, iron can be magnetized, meaning made a magnet, by bringing it in contact with a magnet; the magnet causes a net magnetization in the iron by aligning its dipoles, and it remains even if the magnet is removed. The associated memory effect was used for decades to store information on magnetic tapes. By applying a magnetic field with varying direction it is possible to remove the order between the magnetic dipoles and thereby destroy the net magnetization. This is why magnetic tapes and similar storage media for information must not be exposed to strong, varying magnetic fields; they delete the information.

Last, but not least, magnetic dipole moments can be induced, even if there are no intrinsic magnetic moments. They can be induced in the orbital motion of the electrons, and this is the origin of what is called **diamagnetism**. When an object with charge q, e.g. an electron in an atom, is exposed to a magnetical field B (Fig. 3-22, left), the force F exerted by the field is perpendicular to at least some part of the motion (Eq. 2-54); this leads to a circular motion and therefore an electric current loop. As discussed before (Fig. 2-32), it is itself associated with a magnetic dipole moment. Diamagnetic materials are the magnetic equivalent of dielectric materials.

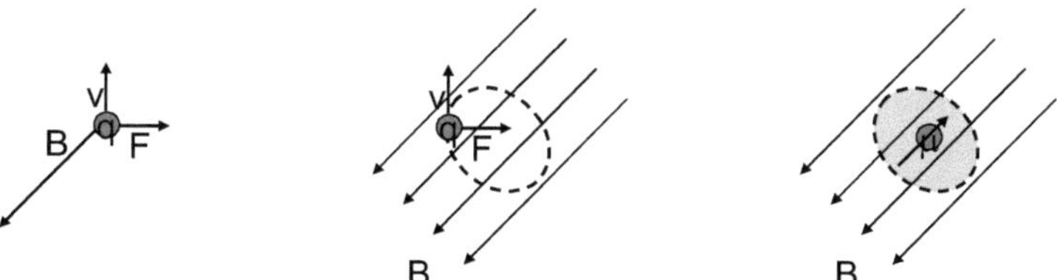

Fig. 3-22 Creation of an electric current loop by an electron moving in a magnetic field (center) and associated magnetic dipole moment (right)

3.1.4.4 Energy conservation and work - energy theorem

There is no need to spend much space and time again to discuss if energy is conserved or not. The same holds for the magnetic interaction as before for the electric interaction and for gravitation: energy is conserved in a single interaction as long as Newton's laws are valid; then, the acting and reacting force in an interaction lead to an action and reaction energy that cancel out. Under these conditions it is clear that energy is also conserved in a situation with many objects. Consequently, the magnetic energy in a system is conserved if there is no effect from the surrounding; if there is one, and energy is exchanged between the system and its surrounding, then energy is conserved taking the system and its surrounding together. There is no need at all to look at the orientation and location of the dipoles in detail.

The same way as before for the electric interaction, we now try to calculate the change of the internal energy when a magnetic field is applied from the surrounding. We already discussed when talking about the internal energy that even the interaction, and thus energy, of two interacting dipoles is difficult to calculate. This is where the work - energy theorem becomes useful.

119

It helps us again to calculate changes of the internal energy, without the need to calculate the internal energy at any specific state. Taking the system and its surrounding together, energy is conserved. Energy can be exchanged between the system and its surrounding: energy is "supplied to" the system, or "retrieved from" the system, depending on the external magnetic field. This is what we call work, and in this case it is called **magnetic work** W_{mag}. How much does the energy in the system change? This is quite similar to the electric case. For a system of many magnetic dipoles the change of the internal magnetic energy is due to the ordering effect of the directed external magnetic field on the magnetic dipole moments within the system. The overall effect on the magnetic dipole moments in a system is called the **magnetization** M of the system. Like for electric energy due to polarization P and an electric field E, for **magnetic energy** associated with a magnetization M and a magnetic field B

$$E_{mag} = \vec{B} \cdot \vec{M} \quad \text{and} \quad dE_{mag} = \vec{B} \cdot d\vec{M}. \qquad \textbf{Eq. 3-13}$$

Again, there is a similarity of Eq. 3-13 and Eq. 2-56, and the opposite sign is again for the same reason as discussed before in section 3.1.3.4.

Summary of the preceding discussion

In contrast to gravitation and the electric interaction, there are no magnetic monopoles. However, the discussion of the magnetic energy of a system of objects, specifically dipoles, showed many similarities to the discussion of electric energy. Regarding external energy, that means energy of the system as a whole, things were practically discussed previously in section 2.2.3.8.: external magnetic energy is due to a net magnetic dipole moment and its orientation in a magnetic field, and in addition the location if the field is inhomogeneous. Regarding internal energy, it is again the sum of the interaction energies of all objects in the system. And regarding the change of the internal energy due to a magnetic field from the surrounding, the situation is practically the same as for electric energy: a direct calculation of absolute energy values from the change of the orientation of the dipoles is complex, often impossible. Employing the work - energy theorem the change of the internal energy can be calculated from the energy change of the surrounding by the applied magnetic field and the change in magnetization by B·dM. Thus, for convenience or lack of detailed information, we usually look again at changes and not at absolute values of the internal energy.

3.1.5 Energy of electric, magnetic and electromagnetic fields

An important effect related to energy, which is not described by Newton's laws, is the energy of electromagnetic fields. Nevertheless, let's make an attempt to explain it starting with Newton's laws and common experience.

Let's first start with the least surprising fact. We already discussed that separating two electric charges of opposite sign, which thus attract each other, is connected with doing work (Fig. 2-25) and thus an increase of the energy of the system consisting of the two separated charges (Fig. 2-26). Because the work done depends on the electric field E (Eq. 2-45), it is obvious that the energy of the system, when calculated, can also be expressed by the field. For a system with two oppositely charged parallel plates, shown Fig. 3-17, this can be calculated rather easy. Such a system is technically a flat plate capacitor, and the electric energy stored in such a capacitor is

$$E_{el} = \frac{1}{2} \cdot Q \cdot U = \frac{1}{2} \cdot C \cdot U^2 , \qquad \textbf{Eq. 3-14}$$

where U is the potential and C the **capacity** of a capacitor, describing its capacity to store a charge by $Q = C \cdot U$. Inserting $C = \varepsilon_0 \cdot A/d$ for a flat plate capacitor and for the field $E = U/d$ (Eq. 3-10) gives (Halliday et al. 1993)

$$E_{el} = \frac{1}{2} \cdot Q \cdot E \cdot d = \frac{1}{2} \cdot \varepsilon_0 \cdot A/d \cdot (E \cdot d)^2 = \frac{1}{2} \cdot \varepsilon_0 \cdot E^2 \cdot A \cdot d . \qquad \textbf{Eq. 3-15}$$

The right side expresses the electric energy using the electric field: the electric energy is the product of the volume and the electric energy per volume. The volume is $A \cdot d$, and thus the electric energy per volume, or electric energy density, is in vacuum $\frac{1}{2} \cdot \varepsilon_0 \cdot E^2$. Expressing the electric energy this way, as **electric field energy**, is for now nothing else than another way of writing down the electric energy, just without the charges and their mutual distance. If the energy is attributed to the electric field, or to the electric charges separated by a distance against a force, is for now a matter of the point of view. Similarly, the **magnetic field energy**, e.g. in a coil or a wire loop due to an electric current, is $\frac{1}{2} \cdot 1/\mu_0 \cdot B^2$. Up to this point, things are not surprising.

To get to something new, let's look next at dynamic situations. Specifically, we look at two dipoles that are interacting (Fig. 3-23) in different situations.

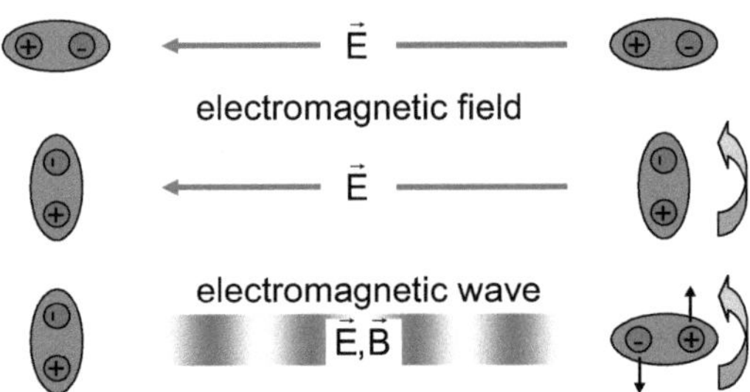

Fig. 3-23 Two dipoles that are interacting; top: static case, center and bottom: dynamic case

The top shows the static case; due to the interaction, which is described by the electric field, both dipoles align. When the right dipole starts to change its orientation by rotating, shown in the center, the left dipole will follow. This is easy if the motions are slow, and still explained by Newton's laws. But if the motions become faster, things start to get complicated. As shown at the bottom, from the viewpoint of the charges in the dipole on the left, the charges in the dipole on the right are an electric current; according to the discussion in section 2.2.3.6 this is the origin of a magnetic field. Then, electric and magnetic field come together. However, this is not everything. If you put your finger in and out of the water in a glass slowly, the water level will rise and then fall; associated with this is a change of the gravitational energy of the water. If you do it fast, you will get a water wave travelling away from the finger. The water goes up and down, thus converting kinetic and potential energy. That water waves carry energy is seen when a wave hits the shore or a boat. An important aspect about waves carrying energy is that they lead us to look only at the wave, and maybe its origin or its destination also, but rarely do we look at the origin and destination that interact by the wave together. The same way, electric and magnetic fields are associated with energy; when charges move fast, there are electromagnetic waves that travel, and they also carry energy. In an **electromagnetic wave**, the electric and magnetic fields cause each other, even without the presence of charges, such that they can travel in vacuum. This is why their velocity c_0 is connected to ε_0 and μ_0. The field, which was introduced as an aid to describe forces in section 2.2.2.2, now itself affects the left dipole; it is "real". An example is light. Electric and magnetic fields, as well as electromagnetic waves carry energy, and are often important in energy balance calculations.

3.1.6 "Nuclear energy"

The nuclear interactions are the last of the fundamental interactions not yet discussed in section 3.1. Their basics were summarized in section 2.2.4, showing that the topic is somewhat hard to grasp because of its development and complexity. Let's see how much we need to understand for complex situations with many objects, and how to deal with it.

Initially, the strong and the weak nuclear interaction were introduced to explain why the protons and neutrons making up the nucleus stick together despite the highly repulsive electric force between the protons, and to explain some nuclear phenomena such as beta decay. The properties of the forces, as derived from the experimental observations, were complex and quite different from the other forces. It was later found that things are much easier.

Today, the weak nuclear interaction is seen as being linked with the electromagnetic interaction; together they are called electroweak interaction. And the real origin of the strong nuclear interaction is found within the nucleons, while what is observed between the nucleons is just a residual effect. The protons and neutrons making up the nucleus, called nucleons, consist of even smaller particles (Fig. 1-1) called quarks, and the real origin is an interaction between the quarks. The interaction between quarks is associated with something like a charge, called color, and because a proton or a neutron consists of three quarks, a proton or neutron thus has a multi-pole field. In addition, quarks have mass and electric charge, and the electromagnetic interaction can be relevant and then another cause for complex behavior of the interaction and correspondingly the related forces between the nucleons. Thus, there are significant differences to the other fundamental interactions.

Up to this point, all fundamental interactions were discussed individually with their associated energy; the discussion in section 2.2 and also here in section 3.1 follows this structure. It is possible with little effort to investigate the interaction and related force between two objects for the cases of gravitation as well as electric and magnetic interaction. For the nuclear interactions this is not the case, and is one reason why it took so long until the interaction between the quarks was discovered. It is hard to investigate; what is commonly observed is the interaction between many quarks, which includes the electromagnetic interaction (gravitation can be neglected). Thus, the nuclear interaction is usually in presence of a second fundamental interaction that is significant; it is practically never observed alone.

It is obvious that this is quite complex, and beyond the scope of this book. However, even in science and technology the treatment is often not detailed. The reason is that a detailed treatment is not needed in practical situations. As discussed already for electric or magnetic energy in complex situations, e.g. materials, science and technology usually try to describe observations. And there is a much simpler way to deal with complex situations with many objects, and also if there is more than one fundamental interaction involved. This way is looking at the energy difference between two states of a system, and knowing that the energy difference is equal to the energy exchanged with the surrounding of the system; up to now we discussed that as work. The work is easier to determine, by the energy change in the surrounding, which can be a system that is less complex. For example, the energy of a nuclear reaction can be determined from the kinetic energy of the resulting particles. And work (and heat) is usually also of our practical interest because it is the needed effort or useful result. Thus, a detailed description of the objects, their motion and interaction, is not necessary. Instead, we can look at the energy difference of the system between two states, described by terms like reaction or binding energy. A **reaction energy** is the difference of the energy of the particles before and after a reaction, while a **binding energy** is the energy needed to separate the particles of a system (the term bond energy refers to a single bond). Now, the term **nuclear energy** is usually used as **nuclear reaction energy**, or as binding energy, and not just as energy of the strong or weak nuclear interaction. This is why it is here, where we discuss energy related to a single fundamental interaction, labeled with quotation marks as "nuclear energy". Despite that reaction and binding energy are energy differences, and allow looking at the energy exchanged between a system and its surrounding, they are still based on the fundamental energy contributions. With enough effort they could, and often are calculated from the difference of the absolute energy of the initial and final state, e.g. in an ionization or chemical reaction (sections 3.2.3 and 3.2.4). Thus, energy conservation can be directly derived from the fundamental energy contributions. But usually they are derived from experimental observations; this is more practical. A violation of energy conservation is never observed.

Nuclear energy is not discussed here any further. The discussion e.g. by reaction and binding energy will be delayed to the following section 3.2, where it is discussed together with chemical energy.

3.1.7 Energy conservation and conversion – more examples

We have now finished discussing the fundamental energy contributions of kinetic energy due to motion, and of gravitational, electric, magnetic, and nuclear energy due to interactions. The term "fundamental" was introduced in this book to signify that they are the basis of all energy contributions / forms, in the same way as the fundamental interactions are the basis of all interactions, including elastic or inelastic collision, and by a shelf etc. Discussing the energy that is associated with each fundamental interaction separately is the easiest way to understand what energy is, why it is conserved, and how it is converted. Section 2.1 started with examples for energy conservation and conversion in simple situations with 1 or 2 objects, and was limited to connections between kinetic and gravitational energy. Section 2.2 included more examples, with additional examples for kinetic and gravitational energy, and examples for electric and gravitational energy. In one example, the connection of work and energy for the case of electric energy was shown. Section 3.1 then included examples for energy conservation and conversion in complex situations, having many objects. Generally, complexity has been increased stepwise, however the discussion of complex situations with many objects is up to now still limited to a single interaction and related energy contribution / form and force. The examples comprise the exchange of energy between a system of objects having kinetic energy interacting by a collision with an object from the surrounding also having kinetic energy, the exchange of energy between a system of objects having gravitational energy by gravitation with an object in its surrounding, and the same for electric and magnetic energy and interaction. These examples significantly extend the concept of energy. Before discussing energy contributions / forms that involve more than one fundamental interaction in the rest of this chapter, it is now a good opportunity to increase the complexity by another step, to add additional insight to the concept of energy.

3.1.7.1 Different energy contribution / form in system and surrounding

In the first example, we discuss a situation with a different energy contribution / form in the system and its surrounding. For this, let's look again at a NaCl crystal, a system of electric charges as discussed before (Fig. 3-13).

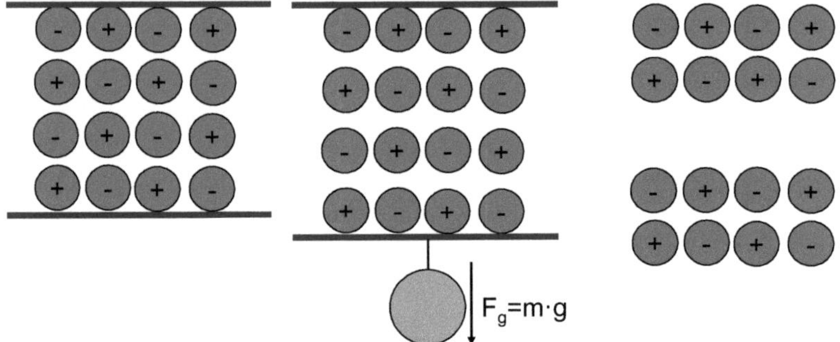

Fig. 3-24 Left: schematic NaCl crystal; center: stretched crystal; right: crystal parts after rupture

Fig. 3-24, on the left, shows the initial crystal, simplified as positive and negative charges. In the middle, the crystal is stretched a bit by the weight of a mass, thus the force of gravitation. Consequently, the distances between the charges in the crystal increase and thus the electric energy of the crystal, its internal energy, increases. At the same time, the mass is lowered such that its gravitational energy decreases. Overall, gravitational energy is used to do work on the system "crystal" by applying the force of gravitation for a distance such that the internal energy of the system "crystal" increases

$$- \Delta E_{sur} = -\Delta E_g = -\int_i^f \vec{F}_g \cdot d\vec{s} = \Delta E_{int} = \Delta E_{el} = \int_i^f \vec{F}_{el} \cdot d\vec{s}. \qquad \textbf{Eq. 3-16}$$

Let's now pull even further until rupture, shown on the right in Fig. 3-24. Rupture means that the distance between the charges at the rupture interface has increased so much that there is no relevant force left. Regarding energy, it means that the electric energy of the crystal has increased, from the low energy level associated with the attractive forces between the charges at the interface where rupture occurred to the high energy level were the charges are at a large distance (Fig. 3-12, Eq. 3-3). The internal energy change is the same as the one from the surrounding used for pulling until rupture occurs. It is like lifting a stone by hand: it gives us a feeling for the force between the stone and the earth, and by the height lifted of the gravitational energy. By pulling until rupture we get a "feeling" for the energy between particles. And it's the same if taking a drop out of a liquid, or a piece out of a solid.

3.1.7.2 Two different energy contributions in the system

The last example had a different energy contribution / form inside than out-side the system, in its surrounding; this is possible by energy conversion. Crucial for the next sections, especially to understand the methods to treat energy related to materials, is to have more than one fundamental energy contribution inside the system, thus an even more complex situation. Let's look again at the example discussed in Fig. 3-17, but as Fig. 3-25 shows, now in addidition to the electric force also with gravitation acting in the sys-tem to keep its two parts together (we can also imagine a spring inbetween).

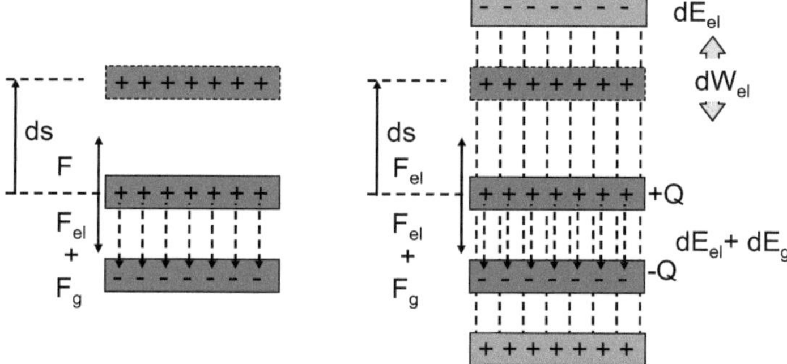

Fig. 3-25 System consisting of two electric charged parts, also subject to gravitation between the parts: left, separated by an unspecified force, and right by an electric force from the surrounding

Again we assume the forces balance each other (static equilibrium), thus

$$dE_{int} = dE_{el} + dE_g = \vec{F}_{el} \cdot d\vec{s} + \vec{F}_g \cdot d\vec{s} = dW = \vec{F} \cdot d\vec{s}. \qquad \textbf{Eq. 3-17}$$

This was easy and straightforward. Now, what if the force and energy from the surrounding are specified, being also electric? Then, Eq. 3-17 becomes

$$dE_{int} = \vec{F}_{el,int} \cdot d\vec{s} + \vec{F}_g \cdot d\vec{s} = dW = \vec{F}_{el,sur} \cdot d\vec{s}. \qquad \textbf{Eq. 3-18}$$

This energy balance can be rewritten, using the polarization, as

$$dE_{int} = \vec{E}_{int} \cdot d\vec{P} + \vec{F}_g \cdot d\vec{s} = dW = dE_{sur} = \vec{E}_{sur} \cdot d\vec{P}. \qquad \textbf{Eq. 3-19}$$

Both sides of the energy balance look similar; however, on one side is the internal created el. field, on the other in the surrounding, and the internal energy has a second energy contribution / form, that of gravitation. Overall, electric energy from the surrounding is used for electric work on the system (electric work as it is by an electric force that acts on charges of the system). The resulting change of the internal energy has two parts: one is electric energy due to the change of the distance between the charges in the system, the another is due to gravitation acting between the charges by their mass. Looking at the example from the concept of forces, the fact that the forces balance and the charge Q is the same means that the electric field outside must be stronger than inside to balance the additional effect of gravitation.

Why is this example important? We often do not know what goes on in a system, especially not in atoms or in materials, which we will discuss soon. We know how energy is exchanged between the system and its surrounding, meaning the type of interaction to exchange energy, e.g. the type of work. Then, the example shows the following crucial things:

1. If we do not have sufficient information to calculate the absolute value of the internal energy, and from it also changes of the internal energy, we can use the work - energy theorem to calculate the changes directly from the work done on the system, more general the energy exchanged. That the result is only the energy change, and not an absolute value, rarely is a drawback; the changes are usually sufficient for applications.

2. The energy exchanged does not tell us completely which energy contributions / forms are present and affected in a system. The example showed that if we use an electric force, we change the electric energy of the system; but we do not know if this is the only energy contribution / form that is present and affected in the system.

3. We also do not know if the energy remains in a certain contribution / form in the system, or if it is internally converted to another one. E.g., if we stir the coffee in a cup, we increase its kinetic energy, but with time it is converted to another energy form.

Thus, energy balances allow calculating a large variety of changes in different situations, without detailed knowledge of what happens within a system. Often, we also do not want to deal with the details. For example, it is possible to choose the energy contribution / form in the surrounding to be one that is easy to use or to measure while the energy contribution / form in the system is not, or unknown.

3.1.8 Summary, conclusions, and further "roadmap"

Summary and conclusions

Chapter 2 discussed energy related to simple situations, which is 1 object in motion, and cases of interaction with a 2nd object by one of the 4 fundamental interactions. For simple situations, dealing just with 1 or 2 objects, it is possible and actually common to describe things using forces, and without talking about energy, its conversion, and conservation. This allows understanding why energy is conserved and how energy is converted. The strength of the concept of energy is however in describing complex situations; so complex that their detailed description is inconvenient, often even impossible, because the situation is not understood in sufficient detail, or because it cannot be calculated with sufficient accuracy.

The goal of chapter 3 is to show that energy conservation and conversion in complex situations with many objects and even more than one fundamental interaction can still be explained and understood, and that conservation can be proved as long as Newton's laws are applicable. Beyond the validity of Newton's laws the derivation does not hold, such that the derivation by Newton's laws is not a general proof. Nevertheless, in many cases it can be argued why energy conservation still holds.

The discussion of complex situations with many objects in chapter 3 is split into two general situations: first, a system of many objects in motion and / or interacting with each other without an effect from the surrounding, and second, with such an effect. The effect is due to interaction with one or many other objects, thus another system, and can lead to an exchange of energy. Thus, the two general, complex situations follow the discussion of the work - energy theorem; however, compared to the discussion in chapter 2 it is now for many objects in motion and / or interacting with each other. Compared to the discussion in chapter 2 (Fig. 2-41), replacing "1 object" now by a "system" of "many objects" (Fig. 3-1) adds another, new aspect. The "1 object" did not change inside e.g. no changes in a ball, the rotating wheel or earth, a dipole etc.; it therefore had no related energy. The system of objects discussed now possibly undergoes internal changes of the position or motion of the objects. Energy associated with internal effects is called internal energy, without such changes we can call it external energy.

Section 3.1 continued the discussion of the fundamental energy contributions from section 2.2; it discussed the cases of kinetic energy, or potential energy due to a single fundamental interaction, for point-like objects as well as for objects with an extension, now for many objects. What are the main results, replacing the previously single object by a system of many objects?

If the system as a whole is in motion or in interaction with its surrounding, it has energy, which we can call **external energy** E_{ext}. The system of many objects can be treated like "1 object" previously, as discussed in section 2.2. It has external energy of the fundamental energy contributions by the motion or position of the system as a whole (Tab. 3-1). External kinetic energy is due to the motion of the whole system by its mass. External gravitational energy is due to gravitational interaction of the whole system by its mass with a mass in its surrounding, represented by a force field or energy field (potential). External electric energy is associated with an electric interaction of the whole system by its charge or dipole moment with a charge or dipole moment in its surrounding, again represented by a force field or energy field (potential). External magnetic energy is associated with a magnetic interaction of the whole system by its dipole moment with a dipole in its surrounding, represented by a force field (there is no simple energy field or potential). External nuclear energy is not relevant as the nuclear interaction only acts at distances that restrict it to the interior of atomic nuclei.

Tab. 3-1 Fundamental energy contributions for complex situations

		energy contribution / form			
		motion	interaction		
		kinetic	gravitational	electric	magnetic	nuclear
E_{ext}	system in motion as a whole	system in interaction as a whole with the surrounding			
E_{int}	objects in the system in motion	objects in the system in mutual interaction			

Energy related to internal effects is called **internal energy** E_{int}. It is the sum of the kinetic energy of the objects in the system due to their motion and the potential energy due to their mutual interaction by the 4 fundamental interactions. Internal kinetic energy is due to the motion of the objects in the system by their mass; it can be changed e.g. by collision with an object from the surrounding. Internal gravitational energy is due to gravitational interaction of the objects of the system by their mass; it can be changed e.g. if the system is in an external gravitational field (not if it is homogeneous); for materials it is not relevant and thus not treated in thermodynamics. Internal electric energy is due to the electric interaction of the objects of the system by their charge or dipole moment; it can be changed by an external electric field. Internal magnetic energy is associated with a magnetic interaction of the objects of the system by their dipole moment; it can be changed by an external magnetic field. Internal nuclear energy does not exist purely; it is accounted for elsewhere (see section 3.2.5).

Energy conservation as well as conversion for all these cases can still be understood from Newton's laws. Without an effect from the surrounding the energy within a system of many objects is conserved. If there is an effect by an interaction, e.g. represented by a force field, energy field, collision etc., then the energy is conserved for the system together with its surrounding. The change of internal energy, which is hard to calculate for complex systems directly, is quite easy to calculate if using the work - energy theorem: for gravitation there is no change of the internal energy dE_{int} due to a gravitation field from the surrounding (unless it is inhomogeneous), for electric energy it is $E_{sur} \cdot dP$, and for magnetic energy it is $B_{sur} \cdot dM$. As the last example showed, if the internal energy comprises more than one energy contribution / form, then the work might affect more than one, and even without any work done internal conversion is possible. This example now leads to the next sections.

Section 3.1 was the continuation of the cases discussed in section 2.2, now for many objects. It discussed the cases of kinetic energy and those with a single fundamental interaction, for point-like objects as well as for objects with an extension. It discussed these cases individually, for two reasons. First, discussing energy contributions due to motion or a single fundamental interaction individually is the easiest way to understand what energy is, why it is conserved, and how it is converted. Second, the fundamental energy contributions cover everything; no part can be missing, and no parts might be counted twice. We could write in continuation of Eq. 2-28 and Eq. 2-57 thus that $E_{tot} = E_{ext} + E_{int}$. This is the perfect situation to write down the energy balance of a situation and use the concept of energy for predictions, optimization etc. But there is a serious problem: the fundamental energy contributions often do not describe directly what we commonly observe, e.g. in science and technology, and this is dealt with in sections 3.2 and 3.3.

131

Further roadmap – revised

The concept of energy must relate to observations that can be made easily, otherwise the concept is too difficult to use, or even useless. Specifically, energy and its conservation correlate changes, and this is only useful if the changes refer to something we can observe. The respective energy contributions are specifically called **energy forms** (section 2.1.4.1). The historic development of the concept of energy started with things easily observable. These are, on a macroscopic scale, the velocity of an object, related to its kinetic energy E_{kin}, and its height above the ground, related to its gravitational energy E_g. Rarely seen, but still observable on a macroscopic scale, is the change of orientation of a compass needle, related to magnetic energy E_{mag}, and similar, fluff sticking to something, related to electric energy E_{el}. But what is the basis of energy released in a chemical reaction, for example combustion? We can see the change in the composition, but not the basis. And where goes the energy if an apple falls and ends on the ground at rest? Here we do not even see what changes. These questions lead to things that we will discuss in the remaining sections of chapter 3: effects on a microscopic scale that we cannot observe by the eye, so their origin was initially not understood. In addition, what we commonly call energy forms usually involve more than one fundamental energy contribution, but the fundamental energy contributions as discussed in section 3.1 are still the whole basis.

The first set of energy forms we discuss is chemical and nuclear energy. Both relate to a change in the composition, in chemical reactions in the electron shells of atoms or molecules, in nuclear reactions in nuclei of atoms. The related energies combine contributions of kinetic energy and of potential energy from the fundamental interactions. There is no "chemical" interaction or force. Their common description is by the energy difference between the initial and final state, called reaction energy. Then, just the amount of substance that reacts, which is observed easily, must be known. No invisible details are needed! The second set of such energy forms is deformation energy and thermal energy. Both relate to the sum of the kinetic and potential energy of all particles of a larger body, made of a material. Deformation energy refers to the change of kinetic and potential energy when the form, meaning size or shape, changes, and thermal energy due to collisions with external particles at the boundary. Moreover, they focus on changes in order and disorder, and together with kinetic and gravitational energy, they are historically where the concepts of work and heat started.

3.2 Energy related to a change in composition

The situation we are looking at now is a system of objects where parts split or combine by loosing or by forming bonds, as shown in Fig. 3-26, again with or without an effect from the surrounding. This situation is treated within the concept of energy quite different to what we have discussed yet. Instead of looking at energy associated with the motion and position of the individual objects by one of the associated fundamental energy contributions, we look at changes we observe and the energy difference associated with that change between the initial and final state of the system.

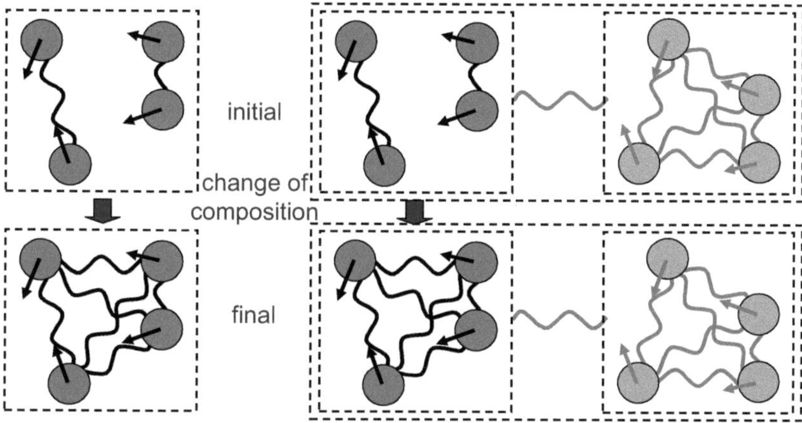

Fig. 3-26 General situations with many objects: a system of objects without (left) and with (right) an effect from the surrounding; now with a change in composition between initial (top) and final (bottom) state

The formation of new particles in a chemical or a nuclear reaction is crucial because of its importance in energy technology, and in the case of chemical reactions also for materials technology. An example is the combination of hydrogen molecules H_2 with oxygen molecules O_2 to water molecules H_2O, called oxidation of hydrogen, or also the reverse reaction. People observed such reactions long before knowing that materials consist of particles, like atoms or molecules, and moreover which interactions act between them. What people observed is the change in composition, and what they could measure is the amount of substance involved. From this, the change in the energy of the system can be calculated by multiplying the amount of substance that reacts with the energy involved when a certain amount reacts. The latter must be determined by other means, and is commonly tabulated. Of course, this way energy conservation is just observed from experiments, and its origin is not understood. Nevertheless, this way is so easy that it is still the commonly used way. Why? We know today atoms, molecules and the interactions between them. The answer is simply: it is so convenient, while the details are extremely complex. What makes it so complex is that we deal with microscopic systems where we do not observe the particles and their interactions, that we have many particles and several fundamental energy contributions involved, and on top that Newton's laws don't apply!

Chemical and nuclear reactions involve e.g. atoms, molecules, or electrons. These particles are so small that we don't see them or their changes directly; we can only observe the change in composition by the change in properties. Strictly speaking, the situation in Fig. 3-26 is not limited to atomic and molecular dimensions; we could also look at an asteroid from the solar system coming into the gravitation field of a planet of the solar system and becoming a new moon. The moon is then in a bound state with the planet. But in macroscopic systems we can observe motion and position, have few objects, usually only kinetic energy and energy due to gravitation, and can even describe things using Newton's laws without the concept of energy. This is not so in microscopic systems, the scale of chemical and nuclear reactions. Chemical and nuclear reactions involve many particles, in macromolecules thousands, and the related energies combine contributions of kinetic energy, and potential energy from a combination of fundamental interactions: electric, magnetic, and nuclear. The energetic effect is due to a combination of fundamental energy contributions and could be described by them in detail; but a description this way would be extremely complex and in addition not related to what we can observe. For example, we do not see the change in the distances between the involved atoms and thereby have no idea of the interaction energy. We do not even see which interactions are involved. What we do observe instead is the change in composition via a change in physical or chemical properties, e.g. H_2 and O_2 are commonly gaseous while H_2O is liquid, and the changes of the amounts of the substances.

But even if we find out what is going on at the microscopic level by using microscopes etc., we cannot describe things by Newton's laws anymore. Not only does the description become complex. The terms that we are so familiar with, position and path of motion, that helped us to understand energy, its conservation, and conversion between energy contributions / forms, they are no more applicable. This means, Newton's laws and what is called Newton mechanics is not applicable; we need to use quantum mechanics.

Before discussing chemical and nuclear reactions and associated energies, it is thus necessary to look at atoms in more detail, especially regarding things affecting energy. This will introduce the necessary basics on quantum mechanics, and show that we can still keep some of the relevant things about energy that we need to understand the concept. As a basis, we will first look at the atom and quantum mechanics and then, as a transition before chemical and nuclear energy, we look at excitation and ionization energy.

3.2.1 The atomic model and quantum mechanics

3.2.1.1 The orbit of the earth around the sun

To understand what goes on in atoms we will first look closer at the orbit of the earth around the sun in the solar system (Fig. 3-27). Which forces act?

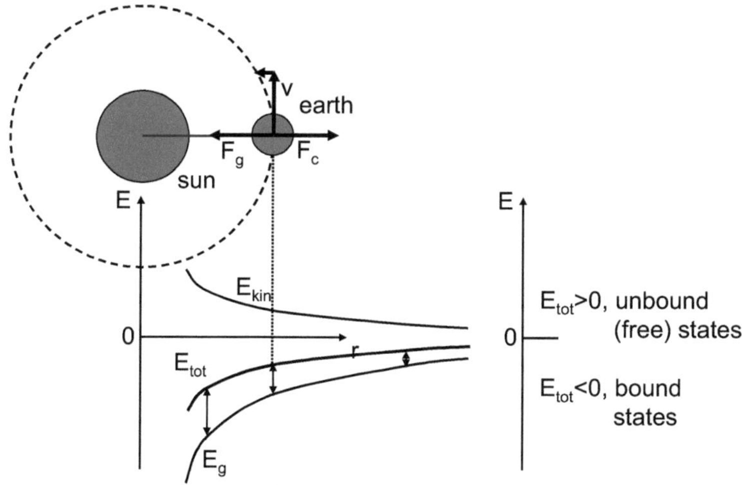

Fig. 3-27 Orbit of the earth around the sun, and associated energies

The situation is comparable to that of swinging a mass on a rope in a circle. Because of its inertia the mass likes to move straight (Newton's 1st law) and thus deviate form the circle outward. The force that drags the mass outward, due to its inertia, is called **centrifugal force** F_c. To make the mass move in a circle, a change of its velocity direction is necessary, and thus an inward force according Newton's 2nd law. This inward force is called **centripetal force**. When swinging a mass on a rope we feel the outward force and balance it by pulling on the rope inward. Instead of pulling on a rope, between the earth and the sun it is the force of gravitation F_g that acts.

For a circular orbit, F_g and F_c must be equal in magnitude and have opposite direction (Fig. 3-27). Both forces depend on the radius r of the orbit: $F_c = m \cdot \omega^2 \cdot r = m \cdot v^2/r$ and $F_g \propto m \cdot 1/r^2$. For a given radius, the gravitational force and thus energy is given (Eq. 2-37, Eq. 2-41). The requirement of both forces being equal in a circular orbit determines the velocity, and consequently also the kinetic energy. The gravitational energy and kinetic energy can then be drawn together in a single energy - radius diagram (Fig. 3-27). The gravitational energy is negative and approaches zero with increasing radius (section 2.2.2, Fig. 2-22), the kinetic energy is positive and decreases with increasing radius, and the total energy E_{tot} (sum of kinetic and potential energy) is negative and increases with increasing radius. Looking at the situation from the viewpoint of energy, there are some important issues:

1. There is more than one energy contribution involved: kinetic energy and gravitational energy.

2. The total energy of the system depends on the distance between the masses, the radius r of the orbit, and it can vary continuously.

3. For $E_{tot} < 0$ the radius is finite; the earth is "bound" to the sun, and the situation is called a **bound state**. For $E_{tot} > 0$ it is unbound.

3.2.1.2 Atomic model and energy levels

What has the orbit of the earth to do with atoms? Atoms are small particles, typically about 10^{-10}m in diameter. An **atom** (Fig. 3-28) is composed of electrons, which have a negative charge, and move around a nucleus, which is positively charged. In an atom, different forces act when an electron moves around the nucleus, similar to the earth that moves around the sun: centrifugal force, and electric force (section 2.2.3) instead of gravitation.

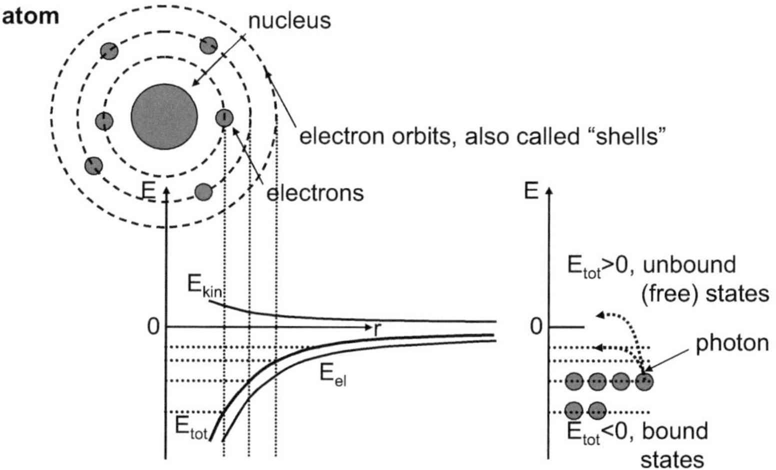

Fig. 3-28 Bohr's atomic model; some allowed levels of the total energy of bound electrons are given separately again on the right

But there are some crucial differences to the orbit of objects around the sun: atoms only absorb or release discrete amounts of energy e.g. by absorbing or emitting photons (Fig. 3-28), and electrons in atoms never fall into the nucleus despite of the attractive electric force. N. Bohr "explained" this in 1913 by a new atomic model that allows only discrete electron orbits and thus energies, and a lowest orbit, called **ground state**, "prohibits" electrons to fall into the nucleus. The discrete energy levels of the electrons are similar to those of books on shelves, with the books having gravitational energy; no book is allowed below the lowest shelf (ground state), e.g. on the floor.

While **Bohr's atomic model** describes the observations, what is the explanation for it? Why are there discrete energy levels? Bohr's explanation was an additional condition besides the one that centrifugal force and electric force balance; it is that the angular momentum m·v·r (section 2.2.1.2) of an electron in an orbit around the nucleus is equal to n·h/2π

$$\Theta \cdot \omega = m \cdot r^2 \cdot \frac{v}{r} = m \cdot v \cdot r = n \cdot h/2\pi, \qquad \textbf{Eq. 3-20}$$

where n is an integer and h is called **Plank's constant**. But why is this so?

3.2.1.3 Quantum mechanics

While in Bohr's atomic model the energy itself is still from a classical model based on forces and Newton's laws, the additional condition that only certain energy levels are allowed is not. It is connected to the idea of a new concept that was developing over centuries: **wave - particle dualism**.

For centuries, there was a discussion if light is a wave or a particle, and it finally changed things far beyond light. The historical process is about as follows (Rooney 2011). As light moves straight in space it seems to be logic that it is a particle; this was what ancient Greek philosopher Empedocles as well as Gassendi and Newton thought. However, others like Hooke and Descartes thought that light is a wave. In 1690 C. Huygens published a theory of light as a wave that explained reflection, refraction, and diffraction. When in 1801 T. Young showed interference of light when passing a double slit, it was clear that light is a wave. In 1817 then, A.-J. Fresnel managed to explain polarization by light being a transverse wave. And J.C. Maxwell's equations of electromagnetic theory, published in 1864, showed what kind of wave: an electromagnetic wave. But there were also observations that are not explained by light being a wave. The first observation was the photoelectric effect, discovered in 1839 by A. Bequerel: only absorption of light of sufficient energy makes an electron in a metal become free to move. The second is the emission of light from a black body. To explain the observations, M. Planck stated in 1901 that light comes in energy packages, which are related to its frequency ν and wavelength λ by

$$E = h \cdot \nu = h \cdot c / \lambda .$$ **Eq. 3-21**

The energy packages are called **quanta**, related to "quantity", hence the name quantum mechanics, and a "light particle" is called **photon**. In 1905, A. Einstein explained the photoelectric effect using Planck's idea of light being made of energy quanta. It was, and still is a disturbing observation and explanation; light behaves like a wave when travelling in space, but is absorbed at a spot with its energy in quanta, like a particle. To make things worse, in 1924 L.-V. de Broglie suggested that the same is true for the particles of matter, and that the additional condition of Bohr (Eq. 3-20) to explain the discrete, stable energy levels in the atom can be explained by standing waves of the electrons in their orbit. A standing wave in the orbit means that the wavelength is a multiple of the circumference of the orbit.

Then, rewriting Eq. 3-20 to m·v = n·h/2πr, and using that the length of the orbit is 2πr and m·v = p, results in what is called the **de Broglie relation**

$$p = h / \lambda,$$
Eq. 3-22

where p is the linear momentum, h is Planck's constant, and λ the wavelength. In 1927, the diffraction of electrons was observed experimentally, which proved their wave nature. These discoveries, specifically the quantization of energy, gave rise to a new way of describing things, which is called **quantum mechanics**, in contrast to **Newton mechanics** that is also called **classical mechanics**.

3.2.1.4 Energy conversion and conservation - systems affected by quantum-mechanics

Due to the scope of this book, it is neither possible nor necessary to discuss quantum mechanics in more detail; such a treatment has e.g. Atkins 1990 and Brown et al. 2015. However, crucial are of course the consequences of quantum mechanics relating to energy conservation and conversion. The applicability of Newton's laws is after all the fundamental basis for understanding the concept of energy from common experience.

Let's first start with the wave - particle dualism. This is disturbing most people about quantum mechanics. The only thing we know is that scientists found ways to describe the observed behavior of photons, electrons etc. Their motion in space, especially when there are obstacles etc., is best described by a wave, so we say they behave like a wave. And when interacting with other matter they can exchange all their energy, so we say they behave like a particle. But that does not mean that they "are" sometimes a wave and sometimes a particle. It means their behavior in different situations is best described by concepts that we are used to from waves and from particles.

To understand energy conservation and conversion it is crucial to remember that quantum mechanics has its roots in Newton mechanics. Bohr's atomic model assumes that the electrons in an atom move in circular orbits around the nucleus, like in a microscopic solar system. The electrons have kinetic energy, and potential energy due to the electric interaction with the charge of the nucleus. Their total energy is then quantized, describing their motion by Newton mechanics plus allowing energy levels only where the angular momentum has a certain value. Of course, this model is very simplified, and quantum mechanical effects have been included only in an improvised way, but it was the starting point to understand atoms, gave some useful results (hydrogen atom), and is still the way of teaching the subject in an introductory course. E.g. Bransden and Joachain 1983 present a calculation and the agreement of the predictions with observations of the energy of photons emitted or absorbed when an electron changes from one orbit to another.

The general mathematical description of a particle in quantum mechanics is by the **Schrödinger equation**

$$H \cdot \psi = E \cdot \psi ,$$

Eq. 3-23

where ψ is the **wave function**. It also derives from its classical description

$$E = E_{kin} + E_{pot} = \frac{p^2}{2 \cdot m} + V ,$$

Eq. 3-24

where $E_{kin} = 1/2 \cdot m \cdot v^2 = p^2/2 \cdot m$ and V stands for the potential energy E_{pot}. Quantum mechanics still uses kinetic and potential energy, the same energy contributions, and even the same relations between the interactions and the energy like Coulomb's law. So what is different, and what isn't? What happens is that by introducing the wave nature the energy becomes quantized, while motion is not anymore described by a path / position. Quantum mechanics does only affect the possible energy levels, but not the energy itself! The effect becomes relevant if the particle wavelength is in the order of the dimension of the available space, which is in microscopic situations. If not, the quantization of energy is too small to be observed, and the energy is correctly described by Newton mechanics. Thus, it does not surprise that the conservation of energy and momentum is still observed and that quantum mechanics does not disagree with Newton mechanics when looking at macroscopic systems. And the observation that energy is conserved in macroscopic situations as well requires that it is conserved in microscopic ones.

The meaning of energy also does not change, nor how energy is converted; conversion is still between the different energy forms that are connected via an object having e.g. mass and electric charge. However, if there is no path or position we cannot use Newton's 3rd law, as we did in section 2.1.3, to derive energy conservation. But already in section 2.1.7 we concluded that at the basis of everything that happens are the interactions, and that forces and thus also what Newton's laws state are just a way of description.

By looking at single interactions, realizing that changes are pairwise with opposite sign such that they cancel out (Fig. 2-18), we see the real basis. The concept of energy describes the effect of interactions between objects; that changes cancel out is a consequence of the definition of energy terms. Thus, we can understand why energy is conserved, but strictly speaking, there is no "thing" that is conserved. Newton's 3rd law is the equivalent in the concept of forces, and there this is obvious. With quantum mechanics in mind, where we have no forces, we could conclude that energy is the more general concept, while the forces and positions in macroscopic situations are the limited area that we can understand from our common experience.

Before continuing, we should realize how important our recent findings are. That energy levels and their differences are not arbitrary any more has an important consequence for the stability of a system, or more precise the state of a system: the discrete energy levels make the state of a system more stable against disturbances. Small perturbations are not possible; only those between the allowed energy states. Long lasting bound states arise, e.g. electrons stay in an orbit around a nucleus and thus atoms become stable. The existence of stable atoms and molecules in general is the consequence, and resulting from it in turn stable, well defined compositions of the elements and (pure) materials. And even more, changes in atoms and molecules are discrete, and also reactions between them. Bound and unbound states are significantly different, and this changes what we observe and use for a classification of two new energy forms: chemical and nuclear energy.

Up to this point, we discussed energy for simple and complex situations, having few or many objects, independent of their size, with and without an effect from the surrounding. This was by looking at the fundamental energy contributions of kinetic energy, gravitational energy, electric and magnetic energy, and finally nuclear energy. We looked at them individually, simply because this is the easiest way to start understanding the concept of energy. Chemical and nuclear energy both relate to several fundamental energy contributions, and changes in otherwise stable particles. The change is in their composition, for nuclear reactions in the nucleus of atoms, and for chemical reactions in the electron shells of atoms or molecules; this means on a very small scale. As the usefulness of the concept of energy is based on making predictions on observable changes, we will now treat things very different from before. To make the transition smoother, let us make two intermediate steps by discussing first excitation energy and ionization energy.

3.2.2 Excitation energy

The different bound states of electrons in atoms (or molecules) are not all occupied by electrons. In a **neutral atom** (or molecule), the number of electrons is equal to the number of protons in the nucleus. Each bound state (Fig. 3-28) can only be occupied by a certain number of electrons because of the **Pauli principle**. The electrons usually occupy the states with the lowest energy levels (Fig. 3-29, left), which are closest to the nucleus. Electrons can move to an unoccupied bound state at higher energy; the atom is then called an **excited atom** (Fig. 3-29, right), the associated energy is called **excitation energy**, and the process is called **excitation**. The reverse process makes an excited atom an unexcited one. Strictly speaking, the rearrangement of the electrons is only a change of composition in a wider sense.

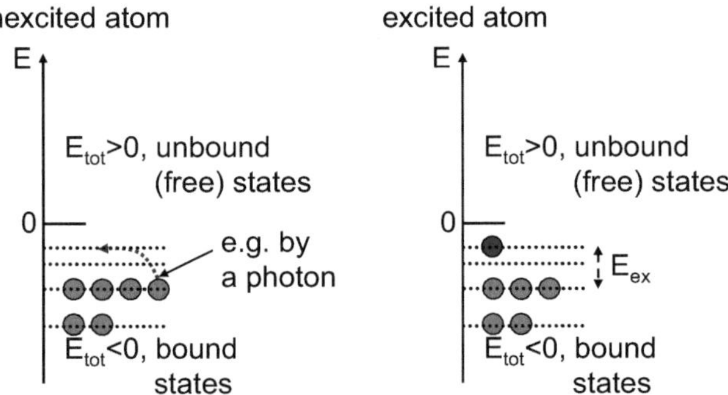

Fig. 3-29 Energy states of an atom: unexcited, left, and excited, right

3.2.2.1 Situations without / with an effect from the surrounding

Fig. 3-29, right, shows the excited atom. Due to one electron being at a higher energy state, it has a higher energy than the unexcited atom where the respective electron is in a lower energy state. Thus, energy is necessary for excitation. Where can it come from? It can come from a collision with another particle like an electron from the surrounding, or from another interaction; due to the charge of the nucleus and the electron this must be an electric, magnetic, or electromagnetic interaction (gravitation and nuclear interaction cannot act on both). Excitation by a collision is common in plasmas, and by electromagnetic interaction in the absorption of a photon.

Now, how do we choose the system that we discuss? The discussion focuses on the change in composition, meaning the formation of different particles from previous ones, including excited atoms. Because our discussion focuses on the basics of energy, a system should be selected in a way that it only exchanges energy. The possible effect from the surrounding is then again an exchange of energy, e.g. absorption of a photon, things we are already familiar with. The excitation of a hydrogen atom, marked by *, is then written as

$$H + h \cdot \nu \rightarrow H^* ; \qquad\qquad\qquad \textbf{Eq. 3-25}$$

the atom is the system, while the photon energy $h \cdot \nu$ is the effect from the surrounding. In the reverse process, a photon is emitted. Because of the discrete energy levels of the atoms, the photon energy is quantized and characteristic for the atom, specifically its element. This results in characteristic wavelengths, and it is what Bohr tried to explain (section 3.2.1.2).

3.2.2.2 External energy without / with an effect from the surrounding

Choosing the system as all particles involved in the change of composition simplifies things. The system, e.g. a neutral atom that is excited, can itself be in motion or exposed to a force field that acts on it as a whole. Wherever these do not affect the process we focus on, the excitation, this can be treated separately in the ways we discussed previously, e.g. in section 3.1. For example, an atom can have kinetic energy as a whole, or as a whole be in a gravitation field, and also an electric or magnetic field as long as they do not affect the excitation process (directly drive the process or indirectly affect it via the energy levels). Thus, there is nothing new to say about external energy when talking about excitation, and consequently we only need to discuss internal energy here.

3.2.2.3 Internal energy without / with an effect from the surrounding

Crucial is now to understand that when talking about a change in composition we talk about energy differences only. This is completely different from the previous discussion. Of course, initial and final states have absolute values of internal energy, and can be described as previously discussed. But we want to avoid that. We do not want to think about the fundamental energy contributions as we do not see the related changes directly, and even several come together and are affected by quantum mechanics. A way of description by what is observed is the **excitation energy** E_{ex} of the system

$$dE_{ex} = E_{ex,1} \cdot dN,$$

Eq. 3-26

where $E_{ex,1}$ is the excitation energy of a single atom (or molecule) as discussed above, and dE_{ex} the energy change due to the number of atoms (or molecules) dN that are excited in a system with many atoms (or molecules).

3.2.2.4 Energy conservation and work - energy theorem

The excitation energy is just the difference between the energies of an initial and final state of an atom (or molecule), which can themselves be traced back to the fundamental energy contributions (section 3.2.1.2). It is thus straightforward from the previous discussion (section 3.2.1.4) to say that energy must be conserved. Excitation is based on the fundamental energy contributions, and there is no reason at all to suspect that energy should not be conserved as quantum mechanics only determines which energies are allowed. A calculation of the excitation energy however is a matter of having detailed information and doing the complex calculations. While the energy levels of the hydrogen atom are the most basic example, the helium atom already has two electrons moving around the nucleus, all interacting with each other, which makes it already a highly complex problem. Thus, instead of calculating the excitation energy, it is again more practical to use the work - energy theorem, e.g. by determining the excitation energy from the energy of photons that are absorbed or emitted in the process.

The work - energy theorem also brings in a new perspective. First, the definition of excitation energy as "Excitation energy E_{ex} is the energy difference between the initial state, the unexcited atom or molecule, and the final state, the excited atom or molecule." is from looking at the energy of a system. From the perspective of the energy exchanged by a system, the definition would be "Excitation energy E_{ex} is the energy needed to move an electron from an unexcited bound state of an atom or molecule to an excited state". For a single atom excited by a photon the excitation energy is $h \cdot v$ by Eq. 3-21. And Eq. 3-26 connects the number of excited atoms with the necessary energy from the surrounding e.g. by the photon energy exchanged with the system. For a number of atoms dN it is then $h \cdot v \cdot dN$.

144

3.2.3 Ionization energy

If a neutral atom (or molecule) looses or gains one or more electrons, it becomes an **ion**, an atom (or molecule) with a positive or negative net charge. The process is called **ionization**. Similar to excitation, ionization refers to a change in the electron shell of an atom (or molecule), but now specifically the number of electrons. A characteristic case is that of an electron that initially occupies a bound state ($E_{tot} < 0$) of a neutral atom (or molecule) and is then removed with just enough energy that it becomes unbound ($E_{tot} = 0$), leaving behind a pos. charged atom (or molecule). For the atom in Fig. 3-28 this shows Fig. 3-30. Another case is a neutral atom (or molecule) that takes up a free, unbound electron into an unoccupied bound state, resulting in a neg. charged atom (or molecule). The energy difference between the initial, neutral atom (or molecule), and the final, ionized atom (or molecule) and the free electron is the **ionization energy** E_{ion}.

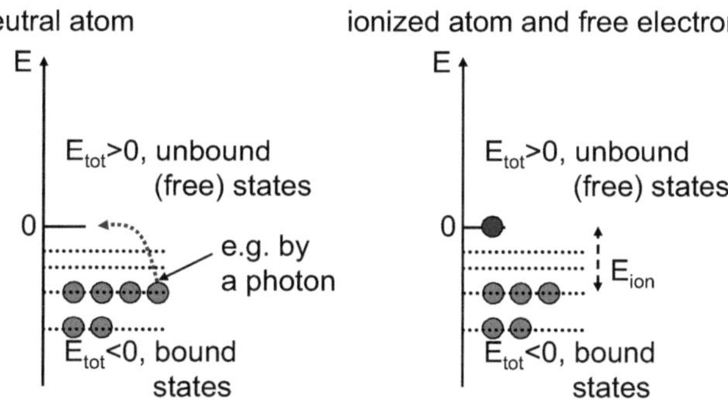

Fig. 3-30 Energy diagram of the initial, neutral atom, and the final, ionized atom and the free electron

3.2.3.1 Situations without / with an effect from the surrounding

Now, what should we choose as the system in the discussion of ionization? As already discussed for excitation, the discussion is again about a change in composition, so we will take again all particles involved in the change of composition as the system. The effect from the surrounding is then again an exchange of energy, e.g. by a collision, or by another interaction that can be represented by a corresponding force field, things that are already familiar.

Fig. 3-30 shows that a free electron has a higher energy than a bound one. This means, for ionization by removing an electron energy is necessary. Where can the energy come from? It can come from the surrounding, by an interaction with another system. Specifically now for the case of ionization, due to the charge of the nucleus and the electron it must be an electric, magnetic, or electromagnetic interaction (gravitation and nuclear interaction cannot act on both), or a direct collision with another particle. Ionization by a collision is common in plasmas, and by a photon (electromagnetic wave) in the photoeffect where an electron is released from a material. For example, the ionization of a hydrogen atom by a photon is written as

$$H + h \cdot \nu \rightarrow H^{+} + e^{-} . \qquad \textbf{Eq. 3-27}$$

The photon energy $h \cdot \nu$ is the effect from the surrounding, while the rest of the equation describes the process of change of composition in the system.

3.2.3.2 External energy without / with an effect from the surrounding

Choosing the system as all particles involved in the change of composition simplifies things. The system, e.g. the neutral atom that is ionized and becomes an ion plus a free electron, can itself be in motion or be exposed to a force field that acts on it as a whole. Wherever these do not affect the process we focus on, the ionization, this can be treated separately in the ways we discussed previously in section 3.1. For example, an atom can have kinetic energy as a whole, or as a whole be in a gravitation field and also an electric or magnetic field, as long as they do not affect the ionization process (directly drive the process or indirectly affect it via the energy levels). Thus, when talking about ionization, there is nothing new that we need to discuss about external energy. This means we only need to discuss internal energy.

3.2.3.3 Internal energy without / with an effect from the surrounding

Like for excitation, initial and final situations of ionization can be described by the fundamental energy contributions, and thus there are absolute values of internal energy. But as before, we don't want to use fundamental energy contributions; the calculations are complex, and we do not see the related changes directly. Again, we like to use a different, direct way of description.

The simple, direct way is just talking about the change of the energy of the system, the **ionization energy** E_{ion} as

$$dE_{ion} = E_{ion,1} \cdot dN,$$

Eq. 3-28

where $E_{ion,1}$ is the ionization energy of the single atom (or molecule) as discussed above, and dE_{ion} is the energy change for a number dN of atoms (or molecules) that are ionized. Eq. 3-28 e.g. connects the observable number of free charges in a substance causing electric conductivity, the change related to the observable change in composition, with the necessary energy exchanged with the system e.g. by the photon energy in the photoelectric effect, or by collisions in a plasma.

3.2.3.4 Energy conservation and work - energy theorem

The ionization energy is just the difference of the energies of an initial and final state of an atom, or a molecule, which can be described by the fundamental energy contributions (section 3.2.1.2). Thus, with regard to energy conservation, it is straightforward from the previous discussion to assume that energy must be conserved. The process is based on the fundamental energy contributions, and there is no reason why energy should not be conserved as quantum mechanics only determines which energies are allowed.

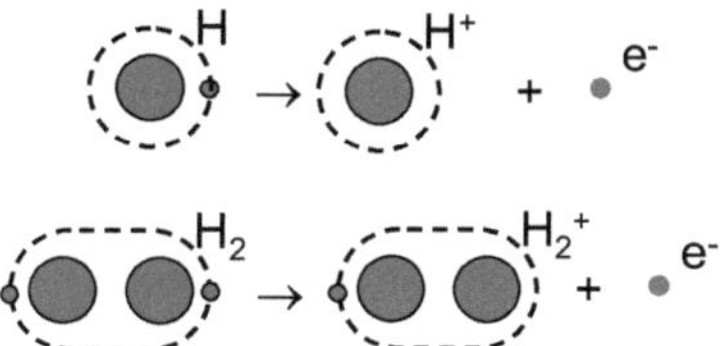

Fig. 3-31 Ionization of a hydrogen atom H, and a hydrogen molecule H$_2$

And the determination of the value of the ionization energy? It is a matter of having all the detailed information, and doing some complex calculations. Fig. 3-31 shows the ionization of a hydrogen atom H and the ionization of a hydrogen molecule H_2. The atomic and molecular structure is just sketched. While the energy levels of the hydrogen atom are the most basic example, the hydrogen molecule is already a quite complex problem. In the hydrogen molecule an electron experiences the electric force of two nuclei such that two potential energies exist, and there are two electrons even interacting with each other; calculated values are thus usually just an approximation.

147

Therefore, instead of calculating the ionization energy from the difference between the calculated energy of the initial and final states, a much easier option is the direct determination of the ionization energy by experiments. Again, this way we take advantage of the fact that the change of the energy of the system is equal to the energy that it exchanges with its surrounding. The definition of ionization energy above "Ionization energy E_{ion} is the energy difference between the initial state, the neutral atom or molecule, and the final state, the ionized atom or molecule and the free electron" looks at the energy of a system. From the perspective of the energy exchanged the definition would be "Ionization energy E_{ion} is the energy needed to remove an electron from a stable bound state of an atom or molecule".

Is the exchanged energy work? The energy needed for the ionization can come from a photon or a collision with an electron from the surrounding. We argued before that these cause a directed effect, so the energy exchanged is work by the definition of force times distance or equivalent (definition in section 2.1.3.4). But this holds only in the classical picture; in quantum mechanics, force and position / path do not exist anymore. Finally, is ionization energy potential energy? It combines two fundamental energy contributions, kinetic and electric energy, so it is not potential energy in a sense that it is a function of position. But it is potential energy in the sense of "hidden" energy that "potentially" can do something.

3.2.4 Chemical energy

After discussing changes of the composition of the electron shell of atoms or molecules, next comes the composition on a larger scale, of materials. **Materials** consist of atoms and molecules, with molecules being themselves the combination of two or more atoms. The field in science that deals with the composition of materials on the level of atoms and molecules, as well as its change, is called **chemistry**, and the respective processes of a change of composition are called **chemical process**. An overview shows Tab. 3-2.

Tab. 3-2 Chemical composition, materials, and processes

composition	material	chemical process
stochiometric	chemical compound	chemical reaction
non-stochiometric	solution / mixture	dissolving / mixing

If a material is composed of atoms of two or more elements in fixed proportions, e.g. NaCl is composed of Na and Cl by 1:1, the composition is called **stochiometric composition**, the material is called **chemical compound**, and a chemical process where the composition of chemical compounds changes is called **chemical reaction**. In contrast, if the composition has variable proportions it is called **non-stochiometric composition**, e.g. a NaCl in water solution, the material is called **solution** or **mixture**, and a corresponding chemical process is called **dissolving** or **mixing**.

What is the reason for the difference between a compound and a solution or mixture? In both cases, atoms and molecules interact and form a **bond**, but the bonds are fundamentally different. They are classified as different bond types. E.g. dipole-dipole bond, ionic bond, and metallic bond are based on the electrostatic interaction; we discussed them before in section 3.1.3.3. Another important bond type is the covalent bond, also called atomic bond.

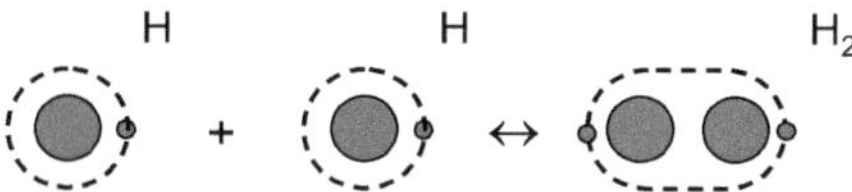

Fig. 3-32 Two hydrogen atoms reacting to a diatomic hydrogen molecule, with electron orbits sketched

In a hydrogen molecule (Fig. 3-32), hydrogen atoms share electrons to get a more stable state. Bonds due to shared electrons are called **covalent bond** or **atomic bond**. They are not just based on the electrostatic interaction. Like in atoms, electrons in electron shells of molecules have kinetic and electric energy, and quantum mechanical effects only allow certain energy states. Sharing electrons makes the compounds energetically more stable.

Now, which bonds lead to a stochiometric and which to a non-stochiometric composition? Sharing electrons in covalent bonds and transferring electrons in ionic bonds is only compatible with a stochiometric composition of the partners; they are thus the basis for chemical compounds. An example for a chemical reaction where new compounds are formed by sharing electrons is the formation of a hydrogen molecule from atomic hydrogen, as shown in Fig. 3-32. The atomic and molecular structure is sketched, indicating the sharing of both electrons in the hydrogen molecule. In short, this is written

$$H + H \longrightarrow H_2 .$$
Eq. 3-29

An example for a stochiometric composition caused by electron transfer in an ionic bond is the reaction of sodium and chlorine to sodium chlorine

$$Na + Cl \longrightarrow Na^+Cl^- .$$ **Eq. 3-30**

Because electrons can only be shared or transferred completely, sharing or transferring them causes fixed proportions, so stochiometric compositions.

The situation is different with the other bond types that are just based on an electrostatic interaction between partners without sharing or transferring electrons. An example is a solution of NaCl in water, consisting of Na^+ and Cl^- ions dissolved in molecules of water H_2O that are dipoles. The bonds between the ions and the water molecules are ion-dipole bonds, and they allow variable proportions; this is expressed by the variable parameter x in

$$NaCl + x \cdot H_2O \leftrightarrow Na^+ + Cl^- + x \cdot H_2O .$$ **Eq. 3-31**

Let's limit the discussion now to chemical reactions and follow the same discussion path as in section 3.1 (solutions are discussed in section 5.4.1.2).

3.2.4.1 Situations without / with an effect from the surrounding

As before, we choose the system as all particles involved in the change of composition (e.g. Eq. 3-29, Eq. 3-30, and Eq. 3-31). This way, no matter leaves or enters the system; only energy is exchanged with the surrounding.

The energy change in a chemical process is called chemical energy. Thus, **chemical energy** refers to a change in the composition of materials on the atomic and molecular level; the energetic effect is due to existing bonds being broken and new ones being formed. For example for the chemical reaction of hydrogen and oxygen to water and reverse, this can be expressed by

$$2 \cdot H_2 + O_2 \leftrightarrow 2 \cdot H_2O + E_{chem} .$$ **Eq. 3-32**

It says that energy is released when water is formed, and needed to split it. This chemical reaction is one of the most promising chemical reactions for future energy applications. The reason is that it involves a considerable amount of chemical energy, which is released by the oxidation of molecular hydrogen to water, and which has to be supplied to reverse the reaction. Technically important is that the reaction can be done in a variety of ways.

For example, the transfer of the necessary energy to split water can be as electric energy from the surrounding; in this case, the reaction is called an **electro-chemical reaction**, specifically electrolysis. It is not surprising that the energy for the reaction can also be supplied by photons; it is then called a **photo-chemical reaction**. Another option is kinetic energy; if the water molecules move fast enough their kinetic energy can supply the needed energy; the reaction is then called a **thermo-chemical reaction**. The connection between kinetic and thermal energy will be discussed in section 3.3.3. When hydrogen and oxygen react back again to water, also different ways can be chosen. Altogether, there is a wide variety of options to choose from for the forward and backward reaction such that it can be adapted to whatever energy is available in the surrounding or needed there.

3.2.4.2 External energy without / with an effect from the surrounding

For the moment, let's just focus on the minimum set of particles needed for a process and see how energy is described in that context, e.g. two hydrogen molecules and a single oxygen molecule that react to two water molecules. The system, the atoms and molecules involved in the chemical process, can be in motion as a whole, or as a whole be in a gravitation field and also an electric or magnetic field. Whenever these do not affect the chemical process this can be treated separately in the ways we discussed in section 3.1. For example, the motion of the system as a whole or a gravitation field does not affect the chemical reaction of the system, the minimum set of particles.

However, when they do affect the chemical process, by directly driving the process or by indirectly affect it via the energy levels, we have to be careful. For example, a very weak electric field would not affect a chemical process, but a stronger electric field could affect the energy levels of the electrons in the electron shells, thereby change the conditions for the chemical process. And if the electric field is strong enough, it can lead to electro-chemical reactions, thus causing a reaction. The same holds if photons cause a reaction. Thus, effects that drive or indirectly affect the chemical process must be accounted for, but then they affect what happens in the system, such that we deal with internal energy. Thus, like before for excitation and ionization, when talking about chemical processes there is nothing new that we need to discuss about external energy. This means we only need to discuss internal energy.

3.2.4.3 Internal energy without / with an effect from the surrounding

The internal energy is the sum of the energy of all particles within a system. It changes from the initial situation to the final situation, and the difference is the energy of the chemical process. The calculation can be done based on the fundamental energy contributions, but as before there is an easier and more practical option to describe the energetic effect. The **chemical energy** E_{chem}, specifically for a chemical reaction called **chemical reaction energy**, can be described by

$$dE_{chem} = \mu \cdot dN ,$$

Eq. 3-33

where μ is the **chemical potential**, which is simply the energy associated with a change between initial and final state of the reaction, for the moment for the minimum set of particles needed for a process, and dN is the number of such sets affected by the process. Eq. 3-33 relates the observable change of an amount of substance, e.g. by appearance, with the necessary energy. The description of the energetic effect here is like for excitation and ionization of atoms or molecules, and very different from before in section 3.1.; there is no reference to what is going on in detail, the forces or interactions involved, what is the cause of the process etc. Chemical processes are described commonly very different, just by the difference of the energy before and after the process per amount of substance, multiplied by the amount of substance that has undergone the chemical process; both can be given referring to the number of particles, or the mass of the material. This way of description is what has been discovered historically, and it is what is still used as it is practical because it is directly related to what is observed: the amount of substance that has changed its composition. And because of that, chemical energy is commonly called an energy form, not just a contribution.

Let's now look at the limitation that we have made, to focus on the minimum set of particles needed for a process. In bulk matter, there are many atoms or molecules and they can be moving, be exposed to electric fields of neighbor molecules … while the bulk is at rest, which means the reaction chamber in the lab. Then, "external energy" of the atoms and molecules is internal energy of the system of many atoms or molecules in bulk matter. The discussion of bulk matter will be done in section 3.3; for the moment, let's note that the above approach Eq. 3-33 for the description still works.

152

3.2.4.4 Energy conservation and work - energy theorem

The energy change in a chemical process is the difference of the energies of the atoms and molecules before and after the process. Bohr's atomic model allows understanding the energy of an atom as energy due to the motion and the position of the charges, based on the fundamental energy contributions. An accurate calculation for atoms and even more for molecules is complex. But to understand why energy is conserved we do not need calculations. The fundamental things don't change: the fundamental energy contributions are the same, also in quantum mechanical calculations. That the position of particles is replaced in quantum mechanics by the wave function only affects the energy differences that are allowed. That energy is conserved in a chemical process is thus rather obvious from what was discussed before. Because of the complexity of chemical processes, and because they go on at the atomic and molecular level, the macroscopic observation in experiments that energy is conserved came first; there was no explanation. Today, we know the atomic and molecular basis, so we can trace it back to the fact that energy is conserved in a single fundamental interaction. This is the basis of energy conservation in general; it cannot be proved, but we can quite well understand it, as discussed in section 2.1.7.

It is not surprising now that energy conservation is the basis for several very useful laws and rules in chemistry. One is that if a process needs a certain amount of energy, the same amount is then released in the reverse process (at the same conditions). Another is **Hess's law** (Atkins 1990): "The overall reaction enthalpy is the sum of the reaction enthalpies of the individual reactions into which a reaction may be divided". **Enthalpy** refers to certain conditions on the bulk matter, something we discuss later in section 3.3.3.7.

Further on, when talking about changes of composition, no matter what type of change and composition, the terms binding energy, bond energy, and reaction energy are frequently used. The **binding energy** is the energy needed to separate the particles of a system, and thus an energy difference between the bound state and the separated state of them at $r = \infty$. The **bond energy** is the same just for a single bond. And finally, the **reaction energy** is the sum of the energies of the bonds that are built minus those separated.

What about the work - energy theorem? The complexity and the microscopic origin of chemical processes is, like with excitation and ionization, the reason why the measurement of the related energy is indirect, from the energy exchanged by the system with its surrounding. As there is no chemical force or work, other types of work are used for this, e.g. electric work. Quite common is also as heat, as we will see later. Today however, being able to do detailed quantum mechanical calculations with fast computers, we can also calculate the energy related to chemical processes directly.

Finally, is chemical energy potential energy? It is due to a mix of kinetic and electric energy, so it is not potential energy as a function of position. However, it can be seen as potential energy in the sense that it is "hidden".

3.2.5 Nuclear energy

Matter is composed of atoms, which consist of a nucleus and electrons in orbits around it (Fig. 1-1). Up to now, we discussed changes of composition by a rearrangement of existing parts or by new parts, just for electron orbits. A rearrangement of an electron to a higher stable orbit in an atom causes the atom to be excited. If an electron leaves an atom, the atom becomes ionized. A third way is sharing electrons in common orbits of atoms or transferring them from the orbit of one atom to that of another in a chemical reaction. Solutions and mixtures, the other type of chemical processes (Tab. 3-2), are not connected with a change in electron orbits. When changing composition the particles don't change; what changes is their relative proportion in bulk. The energetic effect on the internal energy is then just electric energy, as discussed in 3.1.3. However, when looking at a change of composition in materials, solution and mixing processes are grouped into chemical energy. The question of how to group different energy contributions, especially to make sure that an energy balance is correct, will be discussed in chapter 5. Let's discuss now the last option of changes in atoms: changes in the nuclei.

In section 2.2.4, we already discussed the basics of the nuclear interactions in simple situations with 1 or 2 objects. Later, in section 3.1.6, we discussed the case of complex situations with many objects for a single basic energy contribution, and concluded that the nuclear interaction is usually in the presence of a second fundamental interaction, the electric interaction, and that it can be significant. It is thus somewhat similar to that of the processes affecting the electron orbits; it is complex, as it refers to many objects and several fundamental energy contributions, and it is on an even smaller scale. Instead of describing energy changes by looking at all objects and all fundamental energy contributions, we can however also choose an easier way.

The easy way follows what we have already discussed for chemical energy. Energy related to processes in nuclei is commonly called nuclear energy, despite having a nuclear interaction and a related force and energy already. This is why it is labeled with quotation marks as "nuclear energy" when talking about the energy related to the fundamental energy contributions. When speaking of nuclear energy we usually mean the energy of a nuclear reaction, and then not just the energy associated with the nuclear interaction but instead the difference in the energy before and after a nuclear reaction. But before going into details, let's first get an overview of the different situations we talk about.

3.2.5.1 Situations without / with an effect from the surrounding

As nuclear reactions refer to changes in nuclei, let's start looking closer at nuclei. Nuclei are characterized by their number of protons and neutrons; common is however to give the sum of both, thus the number of **nucleons**, and the number of protons. The number of protons is equal to the number of electrons in a neutral atom; it determines its chemical properties and therefore which element an atom with such a **nucleus** belongs to. Atoms with the same number of protons, but different numbers of neutrons in their nucleus, are called **isotope**. E.g. $^{1}_{1}\text{H}$ and $^{2}_{1}\text{H}$ are two different isotopes of hydrogen, both having 1 proton (lower number), but the second also has 1 neutron and thus altogether 2 nucleons (top number). The protons in a nucleus repel each other due to the electrostatic force. The strong (residual) nuclear interaction is however attractive, thus tries to keep all nucleons, protons and neutrons, together (the gravitational force is so small that it can be neglected). Because the strong nuclear interaction acts only at very small distances, nuclei of a certain size with too many protons are in general not stable.

In a **nuclear reaction**, nuclei can combine to new ones, or separate into smaller ones. This is similar to a chemical reaction, but while in a chemical reaction chemical compounds always react to other chemical compounds, the term nuclear reaction refers to the particles within nuclei more general. The splitting of large nuclei is called **nuclear fission**, and the fusion of small nuclei to larger ones is called **nuclear fusion**. In certain cases, both reactions release large amounts of energy; this looks strange as it seems to violate energy conservation, but it only holds for splitting certain large nuclei and fusing certain small nuclei, and not for the same reaction at once.

An example for an initiated nuclear reaction is when a neutron is shot at a uranium nucleus of the isotope $^{235}_{92}U$ thus that the uranium nucleus falls into parts and releases more neutrons. This can cause an avalanche effect. Nuclear power plants use this process, keeping the reaction at constant rate; for this, they need to control the number of neutrons in the nuclear reactor. But the increase of the number of neutrons can also be used intentionally for an avalanche effect; this is the basic working principle of the atomic bomb.

An example of nuclear fusion is the fusion of two hydrogen nuclei, which are just a single proton, to a helium nucleus, which consists of two protons.

Fig. 3-33 Fusion of two hydrogen nuclei to a helium nucleus

This reaction (Fig. 3-33) releases large amounts of nuclear energy, and is what happens in the sun and other stars; predominantly hydrogen fuses to helium, in further stages also to nuclei of other elements.

In the nuclear reactions just discussed the number of neutrons and protons remained constant, but this is not always the case. The neutrons and protons are not the smallest particles in a nucleus. Instead, they do consist of quarks, a neutron of 2 up and 1 down quark, a proton of 1 up and 2 down quarks. By conversion of an up to a down quark an electron and an anti-neutrino are set free (Kube 2009); this is where the weak nuclear interaction is relevant. Hence, a neutron can become a proton plus an electron and an anti-neutrino (Fig. 3-34, left). If this happens in a nucleus of a carbon atom of 6 protons and 8 neutrons (the most common carbon isotope has 6 neutrons), the result is a nucleus of a nitrogen atom with 7 protons and 7 neutrons, and release of an electron (Fig. 3-34, right) and actually also an anti-neutrino. The conversion of a nucleus of one isotope to that of another is called **nuclear decay**, specifically for the release of an electron it is called **β-decay**. It happens spontaneously without an external cause, at a certain probability per time.

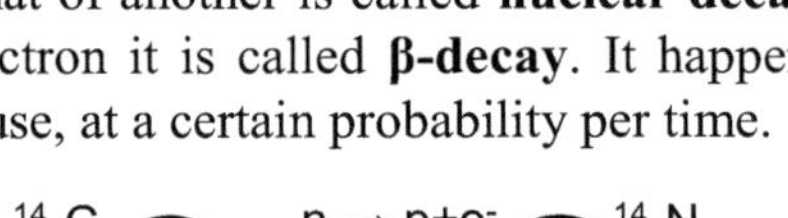

Fig. 3-34 Conversion of a neutron to a proton by conversion of an up to a down quark (left), and its effect in β-decay of a carbon $^{14}_{6}C$ to a nitrogen $^{14}_{7}N$ nucleus

The different types of nuclear reactions are thus fission, fusion, and decay. All are characterized by the existence of different nuclei before and after the process of the nuclear reaction, similar to different chemical compounds in a chemical reaction. In fusion small nuclei combine to a different larger one, in fission a larger one splits into smaller ones, and in decay an internal conversion leads to a new nucleus (without related case in chemical reactions). But like with the electron shell, the reactions are not all types of processes. Nuclei can also be excited, thereby get to a higher energy state. However, there is no process equivalent to ionization.

3.2.5.2 External energy without / with an effect from the surrounding

Looking at external energy without / with an effect from the surrounding, we choose the same approach as before for chemical reactions, ionization, or excitation of atoms: we choose the system as all particles involved in the change of composition, the same way it is also done in process equations. And, as before, we just focus on the minimum set of particles needed for a process, not the bulk, and see how energy is described in that context. Then, the system of particles can have kinetic energy as a whole, or as a whole be in a gravitation, electric, or magnetic field. If they only affect the system as a whole they can be treated as before in section 3.1. There is nothing new to discuss about external energy regarding a change in composition for nuclei.

3.2.5.3 Internal energy without / with an effect from the surrounding

To cause an effect on the composition of a nucleus from the surrounding by a force field, the force field must make a difference on the very short scale where the nuclear forces act. However, neither gravitational, electric, nor magnetic fields from the surrounding are strong enough. The only way the surrounding affects the composition of nuclei is by a direct collision of a particle with the nucleus. An example for an initiated nuclear reaction is when a neutron is shot at a uranium nucleus $^{235}_{92}\text{U}$ and as a result the uranium nucleus falls into parts and releases more neutrons. Then, the initial neutron can be considered as belonging to the surrounding and transferring energy to the uranium nucleus by the collision, or it is seen as a part of the system. In nuclear decay the reaction starts without an effect from the surrounding.

We know now what we have to choose as belonging to the system: all particles involved in the change of composition. The energy of all particles involved, within the system, is then the internal energy: their kinetic energy in the system and their potential energy by mutual interactions in the system. The internal energy changes from the initial situation to the final situation. For a nuclear reaction, the difference is the energy of the nuclear reaction, also called **nuclear energy** E_{nuc}, or **nuclear reaction energy**. We could try to calculate it from the motion of the particles and their mutual interaction, but here the detailed information is simply not available. So, for nuclear reactions, in analogy to chemical reactions (Eq. 3-33), we use simply

$$dE_{nuc} = E_{nuc,1} \cdot dN , \qquad\qquad \textbf{Eq. 3-34}$$

where $E_{nuc,1}$ is the nuclear reaction energy of a single nucleus, and dN is the number of nuclei affected by the process. E.g. for the fusion of hydrogen to helium one could use the energy difference per number of helium atoms resulting from the reaction. Again, Eq. 3-34 connects the observable change in an amount of a substance with the change in the energy related to it. It is just the difference of the energy before and after the reaction per amount of substance, multiplied by the amount of substance that undergoes the nuclear reaction. Optionally we can also use the mass because it is often easier to determine. This description was discovered first, and it is what is still used as it is practical. The reason is that it is directly related to what is observed: the amount of substance that changes its composition. And because of that, nuclear energy is not just an energy contribution, it is what is commonly called an energy form.

Let's now look at the limitation that we have made to focus on the minimum set of particles needed for a process. In bulk matter there are many nuclei as well as maybe free neutrons etc. that take part in nuclear reactions. They can be moving, exposed to electric fields … while the bulk is at rest, e.g. the nuclear fuel in a power plant. Then, external energy of the system of the minimum set of particles is internal energy of the system of bulk matter. The discussion of bulk matter will be done in section 3.3; for the moment, let's note that the above approach Eq. 3-34 for the description still works. This is not surprising as it just states that the energy difference is what is observed per amount of substance that undergoes a change multiplied by the amount that undergoes a change.

3.2.5.4 Energy conservation and work - energy theorem

Nuclear energy is the difference of the energies of an initial and a final state of a nucleus and other particles taking part in a nuclear reaction. This can be described by the fundamental energy contributions, and thus it is straightforward from the previous discussion to assume that energy is conserved. Quantum mechanics only determines which energies are allowed, no more.

However, Einstein's equation

$$E = m \cdot c^2 ,$$
Eq. 3-35

states that mass can become energy, and reverse. There is even a similar way of description for the nuclear binding energy E_b (Kurzweil et al. 2009)

$$E_b = c^2 \cdot \Delta m .$$
Eq. 3-36

It states that the binding energy of the nucleons is equal to the difference in their mass between bound and unbound state, multiplied by c^2; e.g. the binding energy of a proton and a neutron in a deuteron nucleus is $E_b = (m_p + m_n - m_d) \cdot c^2$. In nuclear reactions the mass change can be relevant. This means that energy is not conserved in nuclear reactions, at least not the way we discussed it up to now, and as defined by the scope of this book. According Einstein's equation (Eq. 3-35), and this is what is observed, mass can become energy, and reverse. An explanation of this is beyond the scope of this book. Einstein's equation questions the concepts of particles and mass, which we decided to be a basis for the discussion here. It is quite far away from ordinary experience, thus that it cannot be made understood by a simple example. And its consequences are also far away from common experience thus that it is not necessary to discuss it in detail here. Nevertheless, if this effect is taken into account, we can apply the same way of arguing as before: if the combination of mass and energy is conserved in a single interaction it must also be conserved in several. The corresponding energy is then called **mass - energy** and the law of conservation of energy is truly a **law of conservation of mass - energy** (Halliday et al. 1993).

The exchange of energy with the surrounding we already discussed before. So, finally, is nuclear energy potential energy? Again, it is due to a mix of energy contributions, so it is not potential energy in a sense of a function of position. But it is potential energy in the sense that it is "hidden".

3.2.6 Energy conservation and conversion – more examples

3.2.6.1 Excitation and ionization of atoms

The excitation or ionization of atoms requires an interaction of the atom with its surrounding, e.g. another atom. Both are systems, and colloquially, the interaction is called "exchange" or "transfer" of energy, if there is a relevant distance also "transport". For the interaction two possibilities exist: electromagnetic interaction e.g. by a photon, or collision e.g. with an electron. Gravitation is not an option as it acts on all masses the same way, and the nuclear interaction is not an option as its acts only within a nucleus.

By electromagnetic interaction, described by photons

The electromagnetic interaction acts between electric charges. In atoms, the nucleus has a positive electric charge and the electrons a negative, like a dipole. In section 3.1.5, Fig. 3-23, we already discussed the interaction between two electric dipoles by an electromagnetic field, including as a wave.

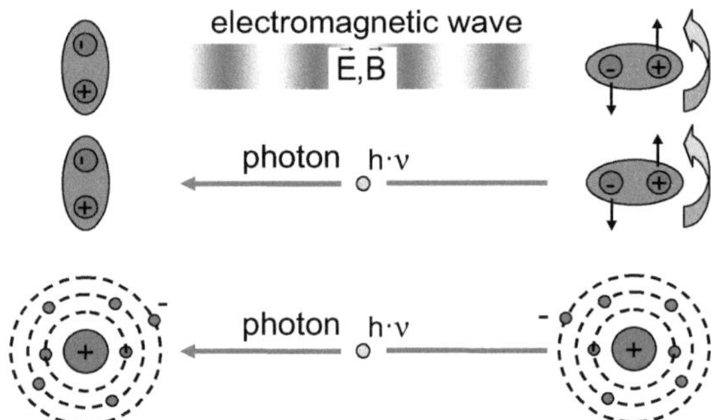

Fig. 3-35 Interaction between two dipoles, and between two atoms, by an electromagnetic wave, and also as described by a photon

We have learned that the electrons in a bound state of an atom have discrete energy levels. Thus, when an electron changes between two bound states, the electromagnetic wave that is emitted or absorbed and carrying the energy can only exchange the energy difference. This is described by a light "particle", called a photon (Fig. 3-35). The energy carried by a photon is related to the wavelength by Eq. 3-21; as a result, wavelength and frequency

160

of the absorbed or emitted electromagnetic wave have discrete values. The existence of these spectral lines was discovered by Fraunhofer in the visible spectrum (light). If unbound states are involved, which have continuous energy levels, then the wavelength can have a continuous spectrum.

By collision, e.g. with an electron

The interaction by collision can for example be with an electron (Fig. 3-36).

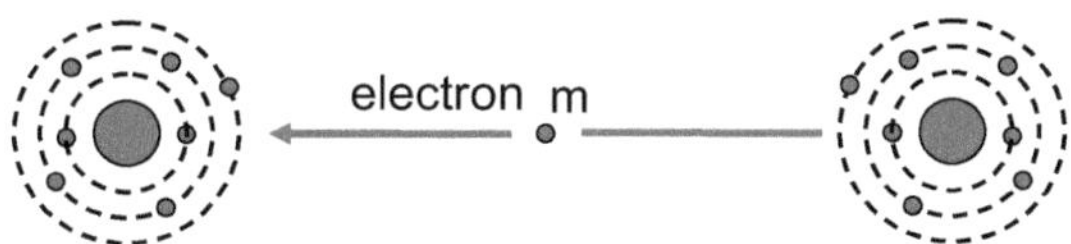

Fig. 3-36 Interaction between two atoms by collision with an electron

An electron colliding with an atom can, if its energy is sufficiently high, kick an electron from a bound state in the atom into an unbound state; the collision is thus inelastic. The atom is then ionized. It can in turn emit a photon if an electron from a higher unbound state occupies the empty place. This is used for the generation of x-rays, which are electromagnetic waves of high energy. For this, electrons are accelerated by an electric field until they have a high kinetic energy, sufficient to kick out electrons of a lower orbit in an atom. The empty place can then be occupied by a free electron, or by an electron from a higher but still bound orbit. Both effects together cause the typical x-ray spectrum, which is continuous with discrete peaks.

In both cases, the interaction by an electromagnetic wave as well as by a collision with an object with mass, there are several common similarities that are in contrast to the cases discussed in section 3.1. First, the excitation and ionization energies are energy differences between two energetic states. Next, the energy of the individual energy states has two contributions from the fundamental energy contributions: kinetic energy and electric energy. Consequently, there is no force or work corresponding to excitation or ionization energies. Instead, the interaction is by the fundamental energy contributions, here by an electromagnetic field or by kinetic energy and collision, which itself needs at least one of the fundamental interactions. After the interaction, the energy that is exchanged ends up as kinetic and (!) electric energy in the atom, similar to the example discussed in section 3.1.7.2.

3.2.6.2 Chemical reactions

To learn more about chemical reactions and energy, let's now look at the reaction of sodium Na and chloride Cl to sodium chloride NaCl. In a simple way this is written Na + Cl $\rightarrow$ NaCl; more detailed, with intermediate steps

$$Na \longrightarrow Na^+ + e^-$$
$$Cl + e^- \longrightarrow Cl^-$$
$$\overline{Na + Cl \longrightarrow NaCl} \ .$$

Eq. 3-37

The reaction is an electron transfer. In a 1st step Na sets an e^- free and is thus ionized; in a 2nd step the e^- is absorbed by Cl, which is thereby ionized too. The resulting Na^+ and Cl^- attract each other due to the electric interaction. The reaction energy is an energy difference between different energetic states; the energy of the individual energy states has two contributions from the fundamental energy contributions: kinetic energy and electric energy. The resulting reaction energy is chemical energy, which is an energy form. Let's first look at the energy in the different situations. We already discussed the energy needed to separate the particles of a system: the energy difference between a bound state and the separated state of $r = \infty$ and $v = 0$. The energy is regained if the system is composed of the separated particles. The separated state, where the interaction vanishes and velocities are zero can be used to normalize the value of energy to zero. This is also possible in this case. Up to now we have just talked about separating a single object. For example, an e^- is separated from an atom, which then becomes an ion. The question is now where to end when separating a compound, or where to start when composing a compound. For the formation of NaCl we could start with Na and Cl, however the natural form of Cl at ambient conditions is diatomic Cl_2. The same way Na in its natural form is not as single atom but in bulk a metal with many atoms interacting by metallic bonds. Further on, NaCl is a salt with many ions interacting by ionic bonds (Fig. 3-13). And this is not even the end. The atoms and ions themselves are composed of electrons and nuclei; the latter in turn are composed of protons and neutrons, which themselves are composed of quarks. The choice of the particles to start with when calculating the energy change to get a compound, or the choice of particles to end with to separate a compound is thus quite large. The calculation of the kinetic and electric energy of a system of particles thus also depends on how deep we look into the system of particles.

For practical reasons, chemists commonly use enthalpy instead of energy; an **enthalpy** difference is a change of internal energy for the bulk at a specific condition, which is constant pressure, as common in lab experiments. The meaning of enthalpy is discussed in detail in section 3.3.3.7; for the moment, we can just think of enthalpy as energy. We already mentioned **Hess's law** in section 3.2.4.4, which is (Atkins 1990) "The overall reaction enthalpy is the sum of the reaction enthalpies of the individual reactions into which a reaction may be divided", a consequence of energy conservation. Let's use Eq. 3-37 as example. Using Hess's law, an unknown reaction enthalpy, e.g. for $Na + Cl \rightarrow NaCl$, can be calculated from the steps it is composed of, e.g. $Na \rightarrow Na^+ + e^-$ and $Cl + e^- \rightarrow Cl^-$, if their reaction enthalpies are known. To standardize this, chemists define the **enthalpy of formation** of a compound as the reaction enthalpy at standard conditions for formation from its elements in their reference states. The reference state of the elements is at the desired temperature and 1 bar pressure, e.g. solid Na and gaseous Cl_2, more precise it is $\frac{1}{2} Cl_2$. From this state, the enthalpy differences to solid NaCl, gaseous Na and Cl, gaseous Na^+ and Cl^- etc. can be found in data tables (Atkins 1990 or Brown et al. 2015).

The case of NaCl is also a good example to discuss energy contributions and energy forms again, and why the difference is so important. One way to look at NaCl is as a result of a chemical reaction; the corresponding energy is then chemical energy, specifically that of a chemical reaction. It refers to the energy difference between the initial and final state of the reaction. However, in section 3.1.3.3 we looked at NaCl as being composed of ions and having internal energy from the electric interaction between the ions. The reaction of Na and Cl to NaCl is thus energetically described by the reaction energy, which is chemical energy and is an energy form, and it can also be described by the different contributions, which is ionization energy in the first step, and then electric energy when the ions finally form NaCl. When making an energy balance we need to pay attention that the individual energy contributions are added up correctly.

We could now repeat the same discussion for nuclear reactions and related reaction energies, binding energies etc., but the discussion would not really add anything relevant that has not been discussed yet.

So, let's now summarize things and draw some conclusions.

3.2.7 Summary and conclusions

Section 3.2 started with an introduction to atoms by Bohr's atomic model. Atoms consist of a positive charged nucleus and negative charged electrons in orbits around it. The energy of an electron in an atom has two contributions: its kinetic energy, and electric energy due to the interaction with the opposite charged nucleus. This is like the earth in its orbit around the sun, but with the electric interaction replacing gravitation. However, experimental observations showed that in an atom only discrete energy levels exist. Newton's laws, treating positions of particles, cannot explain this. However, using quantum mechanics that takes into account the wave nature of particles, instead of Newton mechanics, the correct, quantized energies result. That particles don't have a position is radically different from common experience. However, quantum mechanics is still based on the fundamental energy contributions, especially interactions between mass, charge etc. Thus, despite that Newton's laws cannot be applied anymore, the conservation of energy and conversion between different energy contributions / forms can still be understood. Having only certain energy states instead of continuous ones has two vital consequences: only certain amounts of energy can be exchanged, and as a side effect, there are stable states! Why? If the energy is quantized by a relevant magnitude, then smaller energies cannot disturb and change the system. The fact that the energy difference between states is quantized poses a lower limit. The energies involved in exciting or ionizing an atom, or in a chemical or nuclear reaction, are significantly quantized. To express that two or more particles stay together for a longer time in a stable state, it is commonly said that they are bound to each other by a "bond". In a chemical or a nuclear reaction, bonds are destroyed and / or new ones are formed.

After introducing the atom, energy related to a change in composition was discussed for the cases of excitation and ionization, as well as for chemical and nuclear processes. Excitation is by a transfer of an electron to a higher bound state in an atom or molecule, and ionization to an unbound state. Chemical reactions refer to changes of the composition of chemical compounds by sharing or transferring electrons in electron shells, and nuclear reactions to changes in nuclei. Strictly speaking, nuclei can also be excited, but this is often also classified as a nuclear reaction in a wider sense.

The situations in atoms and molecules involve several objects, more than one fundamental energy contribution, and are quantized. The situations are thus complex and calculations of the energy from the fundamental energy contributions is difficult. But it can be done. However, the energy calculated that way, having several fundamental energy contributions, does not connect to an easily observable change like velocity to kinetic energy, or height to gravitational energy. It is thus not useful in practice. The description is usually using the change in composition directly, observable as different materials with different properties result. For this, there is a general approach to describe the overall energy change: the energy difference between initial and final state for a single process, multiplied by the number of its occurrence e.g. by dN, dm etc. For excitation this is described by Eq. 3-26, for ionization by Eq. 3-28, for chemical reactions by Eq. 3-33, and for nuclear reactions by Eq. 3-34. The calculation provides the energy change, not its absolute value, but the energy difference is usually sufficient for the application of the concept of energy.

Using the energy difference between two states to describe what happens is also still written similar to a force multiplied by a distance, and also the terms generalized force and generalized displacement are usually used. However, this way to describe things is also significantly different to what we discussed in section 3.1. First, in section 3.1 we discussed the value of the internal energy of a state of a system, e.g. gravitational energy or electric energy, and also its change due to an interaction with the surrounding. In contrast, excitation and ionization energies as well as the energies of chemical and nuclear reactions refer directly, thus only, to the energy differences between two energetic states of a system connected by a process. Next, the energy of the individual energy state has more than one fundamental energy contribution. As a consequence, there is no corresponding force or work, at least not directly. There is no exchange of excitation or ionization energy, nor chemical energy or nuclear energy, at least unless mass that carries these energies is transferred across the system boundary. Instead, the interaction is by one of the fundamental energy contributions, e.g. by an electromagnetic field travelling as a wave or by kinetic energy and collision, which itself needs at least one of the fundamental interactions. There are however some cases where the energy exchange can be described by a force or work. For example, in an electro-chemical reaction the electrons taking part in the reaction are forced to move through an electron-conducting wire and to do electric work. Or in the decay of carbon to nitrogen in Fig. 3-34, where the nuclear energy ends up as kinetic energy in a certain direction that could be used to do work. But already two such processes have kinetic energy in different directions. Thus, it is not possible in general to use all the energy exchanged for a force acting a certain distance.

The energy exchange by electromagnetic waves and photons deserves special attention. It is, besides $E = m \cdot c^2$, the crucial deviation from the stage, and therefore quite hard to understand. Local electric or magnetic fields cause a force on a charge and explain thus an energetic interaction well. However, they cannot explain why and how energy is exchanged as quanta. For this, scientists developed another concept, that of exchange particles. The interaction between objects is described as exchange of a "particle", called **exchange particle**, which has momentum and energy, and "explains" the exchange / transfer of them between the interacting objects; it is said to mediate the interaction. This is commonly explained by a ball being thrown between two people forward and backward (causing a repulsive interaction). Electromagnetic waves can be described by such particles, the photons, having energy and momentum. The photons thus explain how energy and momentum is exchanged between atoms etc. as energy quanta. And as there is a time between the emission and the absorption of an electromagnetic wave when exchanging energy e.g. between two atoms, the photon carrying the energy also assures that energy is conserved in interactions at all times. But photons have no mass and no electric charge, so they cannot explain the mechanism of energy transfer. Newton's laws don't apply, and thus the derivation of energy conservation and the simple model do not hold.

The case of NaCl, which can be described as a system of charges with electric energy or as formed by a chemical reaction from Na and Cl has also shown that when making an energy balance we need to pay attention that the individual energy contributions are added up correctly. Specifically the energy differences related to changes of composition are not internal energy in a strict sense as they do not refer to a specific state; they refer to an energy difference between two states, and thus a difference of internal energy.

The examples showed that energy forms like chemical and nuclear energy have developed from practical observation. This leads us to the next section, where we discuss the last set of energy forms: deformation energy and thermal energy.

3.3 Energy related to the relative motion and position

We will now discuss the second set of energy forms that looks at changes we observe and the energy difference associated with that change. The first, discussed in section 3.2, comprises excitation, ionization, chemical, and nuclear energy. The second set, which we discuss now, comprises deformation energy and thermal energy. The changes of the fundamental energy contributions are still on the microscopic level, but not as before within atoms and molecules; they are now in their relative motion and position (Fig. 3-37).

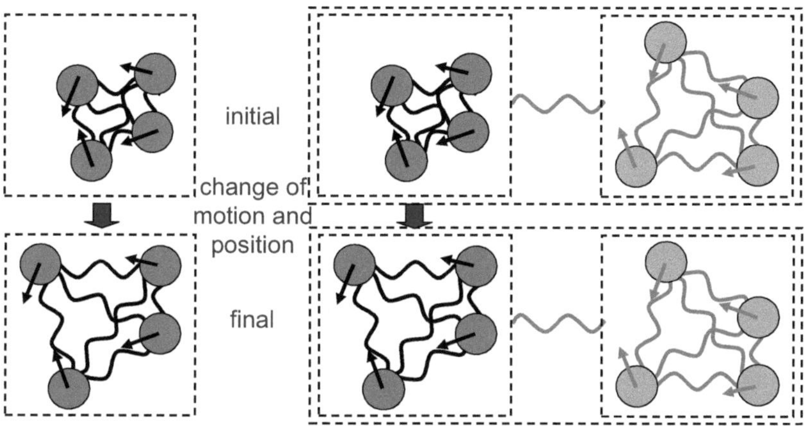

the objects are at rest or in motion (static or dynamic situation), and point-like, or with extension and mass distribution, dipoles ...

Fig. 3-37 General situations with many objects: a system of objects without (left) and with (right) an effect from the surrounding; now with a change of the relative motion and position between top and bottom

With no change in composition, there is no limited number of particles in the system. Instead, we look at very large numbers of particles, typically more than 10^6; thus, we look at macroscopic bodies, or the materials they are composed of. Deformation energy refers to the change of internal energy when the form, meaning size or shape of a macroscopic body changes due to a directed (targeted) force, and thermal energy is due to undirected (random) collisions, e.g. with particles in the surrounding at the surface.

3.3.1 Materials

Macroscopic bodies, which are also objects in the sense discussed before, are made of materials like metals, plastics, water, or air. The term **material** describes matter with a set of well-defined properties throughout, but not size or shape; material is a concept to relate the properties of macroscopic bodies / objects to what they are made of. E.g. the weight of a macroscopic object depends on its size and the density of the material that it is made of. Obviously, changes in macroscopic bodies / objects, and thus in materials, must be integrated into the concept of energy. This was initially done without understanding the microscopic origin, because what materials are made of was not known for a long time. Today we know that they consist of atoms or groups of them called molecules, their ions, and possibly free electrons (Fig. 1-1) in a very large number, and they are in motion and interact with each other. In macroscopic bodies / objects, and thus in materials, there is energy related to motion and interaction of particles on different levels. We already looked at changes within the nuclei of atoms, within the electron shells of atoms and molecules, and within molecules. In section 3.2 we discussed changes within these particles, thus the composition of materials. For this, we looked at few atoms or molecules changing their composition. Let's now look at situations without such changes, thus energy related to the motion and interaction of atoms or molecules, their ions and free electrons. The forces, and related energy and bonds between atoms in molecules, as well as between atoms and molecules in general were discussed in sections 3.1.3.3 and 3.2.4. Their basis is solely the electromagnetic interaction, as gravitation is too weak, and the range of the nuclear interaction is too short.

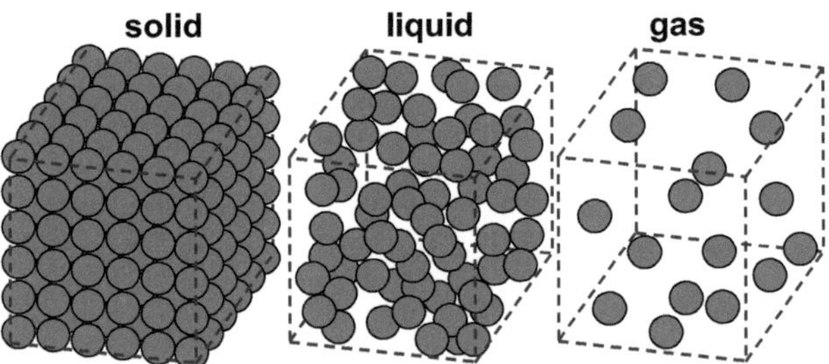

Fig. 3-38 Solids, liquids, and gases are composed of many particles (here spherical particles), which are in motion and interact with each other

Materials are classified as being in a solid, liquid or gas state, as this relates to the most obvious property of a macroscopic body: its shape and changes thereof (Fig. 3-38). In a **solid** the forces between the particles dominate, such that the particles stick together and remain in an average position. The bonds between the particles maintain the shape of a macroscopic body despite the motion of the particles that it is composed of, or any other forces. In a **liquid** the particles also stick together, but even a small force can cause a flow and thus a change of the shape. In a **gas** the particles do not stick together significantly and due to their motion readily fill any empty space or container used to keep them. Because of the ability to change shape easily and thereby flow, liquids and gases are also called fluids. For liquids and gases a containment with walls is needed, which force the liquid or gas to take the shape of the container. This means there is an external effect in any situation at any time. What changes if an additional force is applied?

3.3.2 Deformation energy

Deformation and its associated energy, **deformation energy** E_{def}, refers to changes of the form (hence "deformation") of a macroscopic body exposed to a directed external force or change of force (Fig. 3-39). The term **form** comprises size and shape. Changes of size refer to length and volume, and changes of shape refer to shear, torsion, and bending. For solids all options shown in Fig. 3-39 exist, while for liquids and gases their ability to flow limits the options to a change of size regarding volume, or regarding length if they are contained in a cylinder with piston. It is the macroscopic changes that we observe; we do not see what happens on a microscopic level.

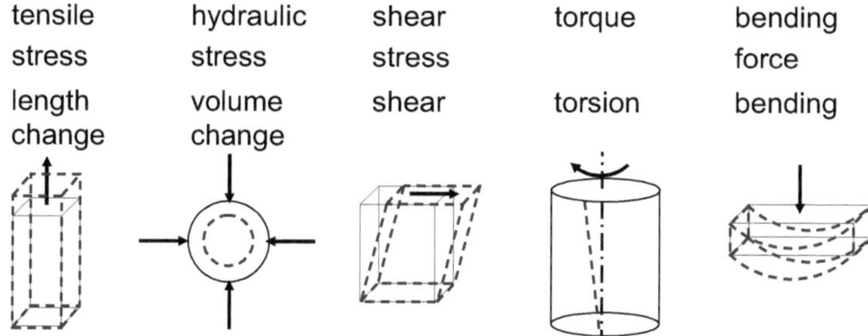

Fig. 3-39 Options for deformation, meaning change of size (length and volume) or shape (shear, torsion, bending)

169

Deformation, like collision, is often accompanied by the term "elastic". That a collision is elastic means that the energy remains completely in the initial energy form, in this case kinetic energy; no loss / gain by conversion to / from another energy contribution / form takes place. In the same way, conversion can also be a cause of inelastic behavior for deformation. Walker et al. 2014 use the term "elastic energy", however it was decided here to use the term "deformation energy" as it is more specific, and we will see in section 3.3.3.5 that it should not be accompanied by "elastic".

3.3.2.1 Situations without / with an effect from the surrounding

The forces associated with the different options of deformation have been used throughout history to change the shape of things, however regarding energy probably the earliest use of deformation is a bow to shoot an arrow. When the bow is bent by hand, work is done on the bow. It stores the energy from pulling as elastic deformation energy. When released, the bow does work on the arrow, giving it kinetic energy to fly far. Wooden, later metallic "bows" were used as shock absorbers in cars, with partly inelastic behavior to dampen oscillations. Energy from tensile stress is related to the length change of slabs; commonly an elastic polymer is used due to the suitable ratio of force to length change. Energy from hydraulic stress is e.g. stored when compressing an air balloon; it is used today in the energy system as Compressed Air Energy Storage, CAES. Energy from elastic shear is stored in metallic springs, and again has widespread applications.

3.3.2.2 External energy without / with an effect from the surrounding

Looking at external energy, the same is true for deformation energy as before for excitation, ionization chemical, and nuclear energy: we choose the system as all particles involved in the change. As shown in Fig. 3-38, this now means many particles in bulk matter, and not just few like in a change of composition. The system of particles can have kinetic energy as a whole, or as a whole be in a gravitation field and also an electric or magnetic field. But as long as these refer to the whole system, and do not affect what we focus on, the deformation, they are already treated previously in section 3.1. Consequently, there is nothing new to discuss about external energy here. Deformation energy is related to systems at rest, and solely internal energy.

3.3.2.3 Internal energy without / with an effect from the surrounding

The macroscopic changes associated with deformation were observed and used long before it was known and accepted that materials consist of atoms. But since we know today that materials are composed of atoms, molecules, ions, and maybe free electrons, and that it is electromagnetic forces that keep them together in materials, the macroscopic observations can be explained and understood.

When the form of a solid body changes, the macroscopic change of size or shape corresponds on a microscopic scale to a change of the relative position of the atoms in the solid body. The relative positions and thereby distance and direction between atoms relate to the interaction between them, and this in turn to the interaction energy. Let's for simplicity look at spherical atoms. If e.g. a solid body of cubic form (Fig. 3-40) is exposed to a force of tension its length changes. On the microscopic scale, the bonds between the particles are stretched in the direction where the force is applied, thus also their average distance increases; the particles relative position changes.

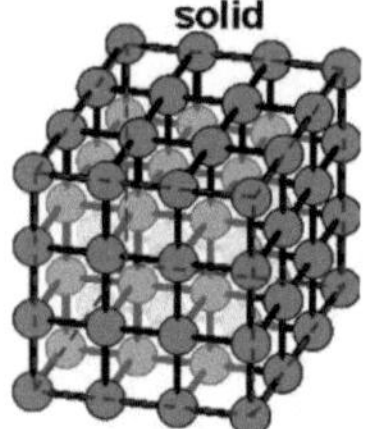

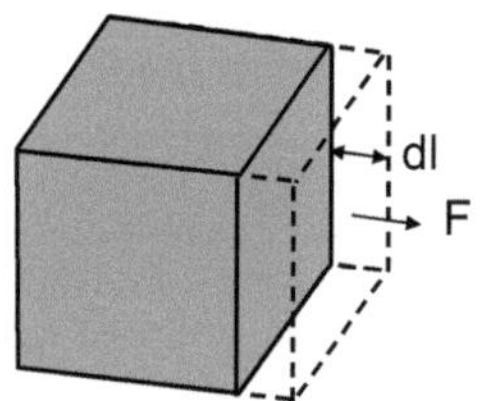

Fig. 3-40 Cubic solid, with spherical particles and bonds indicated by a dash, and deformation by a directed external force changing its length

A directed external force results in an ordered change of the position of the particles in the body. We already discussed this for an ionic crystal in section 3.1.7., Fig. 3-24. Macroscopically, the length change dl caused by the tension force F (Fig. 3-40, left) is associated with **deformation energy** by

$$dE_{def} = F_{def} \cdot dl.$$

Eq. 3-38

Solids always try to get back to their natural size or shape, thus the deformation energy increases when the size de- or increases; to get the correct result in Eq. 3-38 thus "–" has to be used when dl < 0 during compression.

The other options for deformation (Fig. 3-39) are described similarly, e.g.

$$dE_{def} = -p_{def} \cdot dV \qquad \text{Eq. 3-39}$$

describes hydraulic compression. It can be derived from Eq. 3-38 using $p = F/A$ and $dV = A \cdot dl$, with "-" to give an energy increase on compression. The description this way is again by a generalized force and displacement; dl and dV are like dP and dM the macroscopic changes.

What about gases? Gases, like liquids, can be compressed, and Eq. 3-39 that already applies for solids applies again for gases, also including the "-". What about the microscopic basis? Let's look directly at an extreme case, the ideal gas, which is a gas where there is no interaction between particles except collisions. Thus, an ideal gas has only kinetic energy as internal energy, and consequently the relative position is not relevant! But the process of changing the "size" of an ideal gas must affect the kinetic energy. How? When a gas is compressed, the process of moving the boundary walls inward involves collisions with the gas particles (Fig. 3-41) and this increases their velocity; it is like a tennis ball that gets a higher velocity when it is hit. In an ideal gas thus the relative motion changes. Therefore, in general, deformation energy relates to the relative motion and position of particles; the motion or position of the whole system, a macroscopic body, is unchanged.

3.3.2.4 Energy conservation and work - energy theorem

When dealing with deformation and the associated energy, we do for convenience and lack of detailed information usually look at the generalized force and the generalized displacement. For example the energy stored in a compressed spring is $dE_{def} = F \cdot dl$, and if elastic, it is regained when the spring expands again. This way we can observe that energy is conserved, but we cannot understand it because we cannot see its origin. However, we can understand it since we know that materials are composed of particles. From the previous discussion, we know that the internal energy in materials is the sum of the energies in the fundamental energy contributions. As energy is conserved in a single interaction it must also be conserved in many. Quantum mechanics only affects the energy differences that are allowed. So, it doesn't matter if quantum mechanics plays a role or not. It is obvious that energy is conserved without an effect from the surrounding by a force, pressure, or any similar. And if there is such an effect, the energy is conserved as soon as the surrounding is included in the energy balance.

To understand what is going on during deformation, and also with regard to the next topic "thermal energy", we now take a closer look at the microscopic effects. We do this without any interest in the type and origin of the interaction between the particles to avoid unnecessary detail and to get the general idea. So, what happens if a force acts from the surrounding on the boundary, as shown in Fig. 3-41? It acts on the particles at the boundary directly; inside it acts indirectly via inter-particle interactions, incl. collisions. Thus, it is the whole volume that contributes to the overall force and energy.

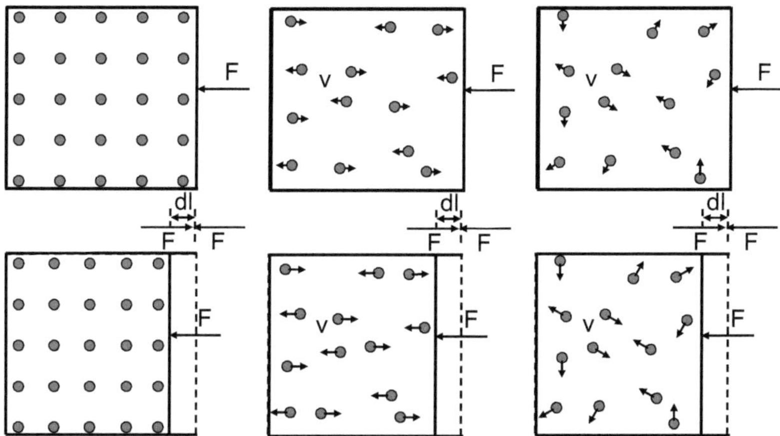

Fig. 3-41 External force acting on a many-particle system, like a solid (left) or a gas (right); the center shows an oversimplified gas

Let's start looking at a solid, sketched in 2 dimensions in Fig. 3-41 on the left. The solid consists of particles located in a regular pattern. The particles are at rest, a simplified case where the particles have no kinetic energy such that we can focus on the potential energy of their interparticle interactions. We now look at the change of the internal energy in three different ways. Macroscopically, if there is a change dl, then Eq. 3-38 with force and length change of the solid body, observed macroscopically, holds. Strictly speaking, it is the force that is acting on the body. But what we measure in an experiment is the force that we apply from the surrounding. We can use the work - energy theorem, and assume that the force on the body is equal to the force applied from the surrounding such that

$$dE_{int} = dE_{def} = F \cdot dl = dW_{def} .$$

Eq. 3-40

173

Microscopically, the force comes from the change of the relative distance of the particles and the interparticle forces due to the interparticle interaction. It is common to visualize them as small springs between the particles. Let's say, as shown on the left in Fig. 3-41, there are 5 times 5 particles in the direction of length change dl; the interparticle distance then changes by dl/4. If the interparticle forces are denoted $F_{i,j}$, the change of the internal energy is

$$dE_{int} = dE_{def} = dE_{pot} = 5 \cdot \left(4 \cdot F_{i,j} \cdot \frac{dl}{4} \right).$$ **Eq. 3-41**

Altogether, Eq. 3-38 and Eq. 3-41 look at deformation energy of the system on the macroscopic and the microscopic level and their connection, while Eq. 3-40 relates them to the energy exchanged as work during deformation.

Let's next look at gases. In 1662, R. Boyle stated from observations that the pressure and the volume of a given amount of gas are related by

$$p \cdot V = const.(T).$$ **Eq. 3-42**

In 1738 then, Bernoulli published an article explaining the behavior of gases by a statistical model with atoms, later called the **kinetic theory of gases**. The model explains that the pressure and thus the internal force on the wall is due to the collisions of the gas particles with the wall, changing their momentum (Eq. 2-2). How often collisions occur per time is given by the number of particles, their velocity, and the wall distance. At a given velocity, halving the volume leads to twice as many collisions with the walls per time, and thus twice the pressure; this also states Eq. 3-42. Strictly speaking, what was just discussed and the statement of Eq. 3-42 hold only for a gas without interparticle interactions, except direct collisions, called **ideal gas**. In a **real gas** there are interactions, but the interparticle distances are large such that interparticle forces are weak; as a result, kinetic energy dominates. Only at close distance, thus at high density, the potential energy is relevant. Generally, in a gas, which is a many-particle system, the particles interact e.g. by collisions, their velocity direction changes and is thus disordered. What changes if a directed force from the surrounding compresses a gas? As Fig. 3-41 shows on the right, a force on a surface, acting a distance dl, just adds another interaction with the particles in contact with the surface. The oversimplified 1-dimensional model in Fig. 3-41, center, where collisions between the particles are neglected, directly shows this. But in reality there are collisions, such that the directed force from the surrounding results in a change of the disordered motion, specifically the velocity (not the disorder itself, as we see later). Upon expansion against a force from the surrounding, collisions with the surface lead to a reduction of the particle velocity in the direction of expansion, and by inter-particle collisions the reduction is compensated by particles with velocities in all directions. Different from the situation in section 3.1.1 (Fig. 3-5, Fig. 3-7), the walls allow using energy from disordered motion for directed work; technically this is realized in a cylinder with a piston, which is the main structure of "thermal" engines. It is striking that "thermal" energy shows up in the discussion of "deformation" energy; we will see soon that this is because both cannot be separated in all cases, especially gases. Let's now look again at the relation between macroscopic and microscopic effects, as done before for solids. While on a microscopic level the situation is complex and difficult to describe, energy conservation allows calculating the energy change of the system by the energy exchanged, by a force acting a distance and thus work

$$dE_{int} = dE_{def} = dW_{def} = F \cdot dl = F/A \cdot A \cdot dl = -p \cdot dV . \qquad \textbf{Eq. 3-43}$$

The change of internal energy can thus be calculated without knowing anything about the effects going on at the microscopic level, not even if the gas is an ideal gas or a real gas. However, for the simple situation of an ideal gas where there is only kinetic energy we can easily write it down

$$dE_{int} = dE_{def} = dE_{kin} = \sum_i \frac{1}{2} \cdot m \cdot d\left(v_i^2\right). \qquad \textbf{Eq. 3-44}$$

Actually, for the ideal gas we could even write down the absolute value of the internal energy, just as the sum of the kinetic energy of the particles.

We have now discussed deformation energy of a solid for the simple case where the internal energy was only potential energy, and of a gas for the simple case where the internal energy was only kinetic energy. In general, kinetic energy and potential energy are present in solids, liquids, and gases. Deformation energy is thus the energy associated with the deformation of a solid, liquid, or gas body by a directed force on that body. Microscopically, deformation is associated with a change of the relative motion and position of the particles, thus deformation energy is associated with a change of the kinetic energy and potential energy due to particle interactions.

3.3.3 Thermal energy

We have now discussed all energy forms, except thermal energy. It will bring the answer to some still open questions, and add some new aspects.

Let's start looking at something we already discussed: an object, for example a soccer or a tennis ball, which falls free (section 2.1.4). From the conservation of energy we concluded that the sum of kinetic and gravitational energy is constant (Eq. 2-28). A new, interesting observation comes when, after bouncing a few times, the ball finally comes to rest. Its kinetic energy is zero again as before the fall, and its gravitational energy has decreased. Where did the energy go? Similar observations can be made everywhere. That energy of motion of a large object decreases and finally becomes zero unless there is a continued driving force is what commonly is observed, for example with a bike, a car, or a plane. Already Galilei explained this by the presence of an effect called **friction**, however without knowing its cause. Friction can be associated with a force; it explains why the motion ceases if there is no continued driving force. The force of friction is often called a **non-conservative force** as it seems as if energy conservation is violated. But the previous discussion has shown that energy conservation is reliable, and looking at the fundamental energy contributions we also understand it. If energy seems to be not conserved it is clear that it just became invisible. But how? The friction of air that slows down the fall of a feather does not give an immediate hint. But friction is present in many situations. For example, when we rub our hands they get warm; thus friction leads to a rise of the temperature. The same was observed when boring holes in metal to produce cannons, and people noticed that this means that mechanical work produced what is commonly called heat. Then, the first steam engine, developed by D. Papin in 1690, showed that heat can also be used to get work. In a thermal engine, a gas is heated, and then it is let to expand to do work. The expansion is what we have discussed as "deformation". And "heating"? If heat can do work, is heat then related to energy in the same way as work? R. Clausius answered the question, stating the **1st law of thermodynamics** as **"the change of the internal energy of a system is equal to the amount of heat supplied to the system, and less the work done by the system"**. Thus, **heat and work denote energy exchanged between a system and its surrounding**. And "heating" and "working" are the associated processes. But why do we need a second term for the exchange of energy besides work? And what is the energy contribution / form that we are still missing?

We concluded in section 3.1.1.4 that friction, leading to an invisible energy contribution / form, is connected with motion on a microscopic level. Now, as **work** is energy exchanged by a directed force acting a distance, what about undirected, random forces? When many particles on a microscopic level interact, they can interact in all directions. Thus, in contrast to work, **heat** is energy exchanged by undirected, random forces on a microscopic level. And undirected, random forces acting on a system of many microscopic particles must result in an undirected motion of them. Due to its microscopic nature, we cannot see this motion, so energy seems to disappear. The first time microscopic motion was observed was in Brown's experiment where small bodies, visible in the microscope, were pushed around by still invisible atoms and molecules. The bodies we can see directly with the eye are not pushed around enough that we are able to see them moving around. But there are many other situations where the random motion of microscopic particles is involved and where we can understand its true nature. What we need is observations where it pushes around small particles, and where we can observe their motion without seeing them directly. Similar to **Brownian motion**, it is observed by the diffusion of a substance in another. In **diffusion**, a substance that is immersed in another substance moves by random motion, for example it starts initially in a small area and after some time is found in a wider area. The reason for the random motion is collisions of particles of the substance with particles of the substance it is immersed in due to their microscopic random motion. Visible, but not by the individual particles, is diffusion of a dye in water; no microscope is needed. Diffusion works without stirring and thus without any macroscopic motion; it happens by microscopic motion. The particles of the dye are pushed around by the particles of the water, and the resulting random motion distributes the dye from its initial location into a wide area. Diffusion is faster the higher the temperature. This shows something important: **temperature** is in some way a measure for the microscopic random motion, which is why it is also called **thermal motion**. If we wait long enough, the thermal motion in a system, e.g. a glass of water, must be the same everywhere even if it was initially not that way. The reason is simply the collisions of the particles of the system with each other. Thus, the temperature must also be the same everywhere, which is called **thermal equilibrium**. Energy related to thermal motion is consequently called **thermal energy**. The temperature obviously takes the place of the generalized force, e.g. like the pressure p. But what is the change that corresponds to the volume change dV?

3.3.3.1 Situations without / with an effect from the surrounding

The deformation of a solid by a directed force from the surrounding acting a distance, and thus by work, is shown in Fig. 3-41 on the left; here, the particles have no kinetic energy, only potential energy due to their interactions. The deformation is associated with a directed relocation of the particles, the average position in the direction of the force from the surrounding, and thus potential energy. What about heat and thermal energy? The exchange of energy at the boundary of a solid by collisions with particles shows Fig. 3-42. Because the motion is random it is not work; it is instead what we call **heat**.

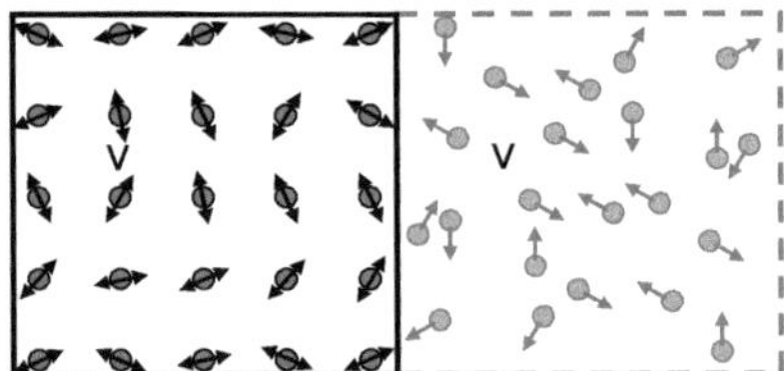

Fig. 3-42 Solid (left) exchanging energy at its boundary by random collisions with particles from the surrounding (right), thus as heat

As a result, the particles in the solid move randomly. This was first observed 1827 as Brownian motion. In 1905 Einstein derived a mathematical model to describe it, and in 1907 a microscopic model of the **heat capacity** of solids, which relates the amount of energy transferred as heat to a solid and the related change of the temperature. The **Einstein model** assumes that the basis is oscillations of individual atoms around their average position and takes into account quantum mechanical effects. The oscillation is due to each atom interacting with its neighbors via interatomic bonds, and thus involves kinetic as well as potential energy, like in a pendulum but with the interatomic bonds replacing gravitation.

Thermal energy is related to the random motion of the atoms around their average positions, and thus potential as well as kinetic energy. However, is thermal energy equal (!) to the energy of the random motion of the particles, as is often said? This would give it an absolute value, which is in contradiction to some things. First, deformation energy is the energy associated with a change of form, e.g. by length or volume, and thermal energy should also be defined by some change. Another hint comes from thermal expansion.

We can from the previous discussion understand the **thermal expansion** of solids as follows: the wider the particles swing, the larger the overall dimension of a solid if its volume is not restricted. Thermal motion affects the force needed to deform a solid. This indicates that there is a connection between thermal energy and deformation energy. We can see it in an ideal gas. In an ideal gas, there is no interaction between particles besides collisions. The internal energy comprises thus only kinetic energy due to the particle motion, and no potential energy. As there is only a single energy contribution, deformation and thermal energy must both relate to the motion of the particles somehow, and both as a change of it. The deformation of an ideal gas by a directed force from the surrounding acting a distance, and thus by work, shows Fig. 3-41 on the right; the particles have only kinetic energy. Deformation means a change of the relative distance of particles, however this only relates to potential energy if there are interparticle interactions (besides collisions). An ideal gas has only energy of random motion, thus kinetic energy. During deformation, the collisions with the moving container walls change the velocity of the particles; however, they do not affect the randomness of the motion. What about heat and thermal energy?

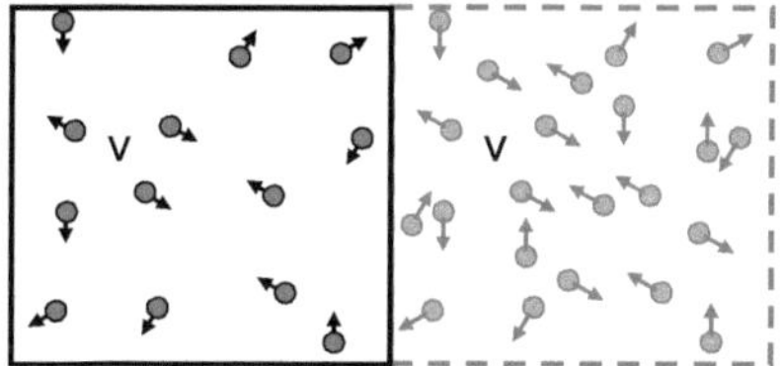

Fig. 3-43 Gas in a container (left) exchanging energy by random collisions with particles hitting the container (right), thus as heat

Fig. 3-43 shows the exchange of energy at a boundary by random collisions with other particles, thus as **heat**. The undirected, random forces acting on the system of many microscopic particles result in undirected motion of the particles; they affect the velocity and the randomness of the motion.

In general, the situations can be much more complicated than we discussed. Real gases have interactions between the particles, thus potential energy, and if the particles are molecules there is also intrinsic kinetic and potential energy by vibration of molecule parts and rotation of the whole molecule. For liquids the situation is so complex that we will not even try to discuss it. But this is also not necessary.

Let's summarize the most crucial things that we found. Fundamental is that the randomness plays a crucial role in distinguishing the exchange of energy as well as in the energy contributions / forms in a system. Work and heat refer to an exchange of energy between a system and its surrounding. Work as a directed force acting a distance does not affect the randomness of motion and interaction in the system, while heat does as it is by microscopic random interaction at the system boundary. And energy contributions / forms? Work done on a system, by a directed force acting on the system, leads to a change in its form and thereby the average distance of the particles in it. The energy associated with the deformation in a solid is associated with the change of the potential energy by the interparticle forces, in an ideal gas it is a change of the kinetic energy. In no case does it change the randomness. Even if the kinetic energy in an ideal gas changes by deformation, its randomness does not change; only the velocities of the particles are changing. Heat supplied to a system, by the random interactions at the surface, changes its thermal energy. In a solid this is associated with a change of the random motion of the particles around their average position, thus like in a pendulum with kinetic as well as potential energy, including the randomness of this motion. In an ideal gas the internal energy only comprises kinetic energy of the particles, which is always from motion that is random by the collisions within the system. Thus, heating by undirected, random forces acting on the system at its boundary changes the randomness of the motion, not just the velocities of the particles. Thus, thermal energy is not equal to the energy of the random motion of the particles.

The energy forms of deformation and thermal energy thus refer to the energy associated with a change, for deformation of length or volume etc., and for thermal energy of the randomness of the motion. This is significantly different from the energy forms discussed in section 3.1 "Energy related to a single fundamental energy contribution", where we could quantify the energy of a system from its state by the motion and position of the objects, and section 3.2 "Energy related to a change in composition", where we could quantify the energy as the difference between two states of composition.

But this is not everything yet. If deformation is not elastic, work done for deformation can end up in other energy forms than deformation energy (e.g. example in section 3.1.7.1). In a similar way, thermal energy can increase even if there is no heat transfer, e.g. by friction, thus by conversion. To get a better understanding of thermal energy, let's look closer at friction now.

The connection of friction, thermal motion, and energy is by several ways. Fig. 3-44 shows a larger particle interacting with a stream of small particles; from left to right: initial situation, after few collisions, and later state.

Fig. 3-44 Larger particle interacting with a stream of small particles. From left to right: initial situation, after few collisions, later state

The collisions lead to a loss of kinetic energy of the larger particle from its directed motion, while the small particles gain kinetic energy in undirected motion. Macroscopically, this loss is by conversion and called **friction**. Friction was already used by Galilei to explain that objects should ideally fall the same way, despite that observations showed that they fall different. Galilei correctly explained that friction is caused by the surrounding of the object and not inherent in the object, and stated that if friction could be eliminated from an experiment then objects would fall the same way. Knowing today that air consists of particles, the air molecules, we have the microscopic explanation. Feathers and chunks of lead fall different because feathers have a large surface and thus they collide with more air molecules; and the collisions slow down the fall of feathers. If a feather is compressed it falls faster, and the same a piece of paper. And if we add more surface, like with a parachute, we fall slower. With airplane models, friction by air is often visualized by adding smoke particles to the air, showing the convection eddies produced by the plane model in the air stream. With time, the eddies become smaller and finally become microscopic, not visible at all. The directed motion of the large particle or the airplane is reduced, while the undirected motion of the many small particles increases; the kinetic energy of the large particle, easily visible, is "lost" (exchanged) and ends up as invisible thermal motion. The same happens in liquids when ships move, or when stirring a coup of coffee. Liquids and gases cause comparatively small friction forces, thus little increase of thermal energy and temperature.

The effect is bigger between solids (Fig. 3-45), especially if pressed against each other to optimize the particle interaction at the surfaces. For example, moving two pieces of wood past each other can been used to start a fire.

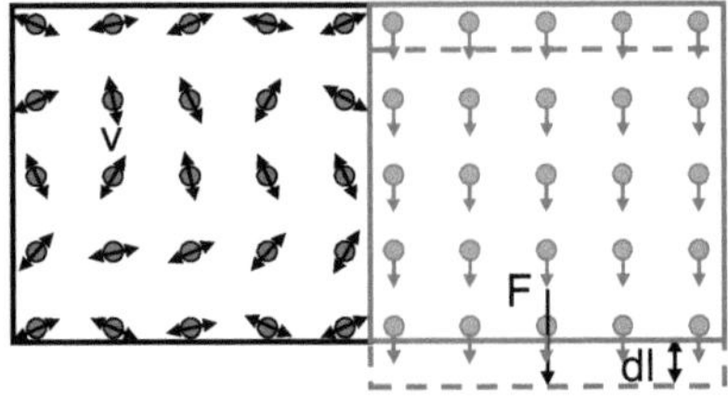

Fig. 3-45 Solids moving past each other; friction reduces the energy of the one with directed motion (right) and increases the thermal energy of the other (left)

While for deformation there is a directed force that moves the boundary, now there is a directed force and motion along the boundary, while dV = 0. When the solids are pressed against each other, the atoms and molecules at the interface interact and exchange energy. As a result, the directed motion of the one that moves is hindered, as experienced by the force of friction opposing its movement, while the thermal motion in the other one increases. We can observe the effect on our body: when we rub our hands, friction leads to an increase of the thermal energy in the skin and its temperature. The same was observed when drilling the channel for bullets into a cannon. Today, we can observe these things more detailed using an infrared camera.

But there is also a way to increase the thermal motion without heat or work, therefore completely without exchanging any energy with the surrounding. Let's look at gas particles in a closed container, and their motion with time. Fig. 3-46 shows two optional initial situations and a single final situation.

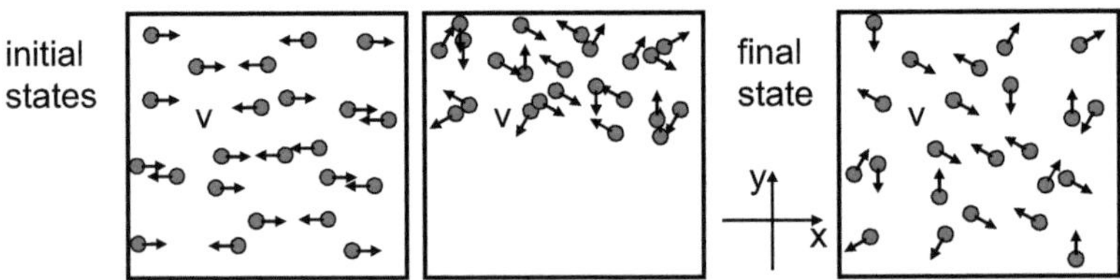

Fig. 3-46 Container filled with a gas consisting of many gas particles; their position and motion changes with time, increasing disorder

Even if the gas particles initially move in only one direction, the x-direction, the direction will become a distribution due to interactions with other gas particles by collisions, and finally will be completely random. Neither the total energy nor the total momentum changes in the process. And if they are initially in one half of the container, they are finally distributed everywhere. Even if the velocity is initially identical, it is finally a velocity distribution. No matter what the ordered initial situation is, interactions will lead to the same final disordered state. Many-particle systems with interactions between the particles always tend to develop towards a more disordered state. Because this determines in general the direction in which systems and their properties change, a new term was introduced as a measure for the disorder: **entropy**, S. That the entropy can increase but never decrease in an isolated system (of many particles) is expressed by the **2nd law of thermodynamics**

$$dS \geq 0.$$
Eq. 3-45

For spontaneous changes, the ">" holds. Now, what is the basis of entropy? For 2 particles there is one option that the particles move the same direction, and one they move opposite. With 4 particles there is only one option that they move the same way, and already several that they move opposite. For millions of particles the probability of the highly ordered state is negligible. And the same holds also for the distribution of their velocities and positions. It is not impossible that a disordered state develops into an ordered state again, however the higher the number of particles, the more situations exist that are disordered, and the lower the probability that the system develops back to an ordered state. This is simply by statistics, and the reason why the description of systems with disorder is called **statistical thermodynamics**. A derivation of how entropy is related in detail to the motion and position of the individual particles is beyond the scope here. However, generally

$$S = k \cdot \ln \Omega,$$
Eq. 3-46

with k being Boltzmann's constant and Ω the number of accessible states, meaning the number of states possible at given boundary conditions, for example in an isolated system and thus at a given total energy.

Disorder tends to increase automatically if a system is isolated, as just discussed, but as discussed before it can also come from the surface in two ways. The direct case, when there is thermal motion outside the system and energy exchange is by collisions at the surface, thus by heat transfer, is one.

The system is then said to be in **thermal contact** with its surrounding. If the thermal motion of the particles inside and outside is the same, nothing changes and the system is said to be in **thermal equilibrium** with its surrounding; both have the same temperature. If the temperature in the system is lower than that of the surrounding energy "flows" as heat into the system, by interactions between atoms and molecules at the boundary, and the thermal motion in the system increases. If its temperature is higher, energy "flows" as heat out of the system and its thermal motion decreases. Then, the system can also loose disorder and thus entropy, but taking the system and its surrounding together the disorder and thus entropy still increases. This can be understood easily thinking of two systems of the same content in thermal contact. If they initially have different temperatures, heat will flow between them until their temperature is the same; in the process the entropy of one system goes down while it goes up in the other. But thinking of both systems together as a single system the discussion before shows that the overall disorder and thus entropy must increase. The same is of course the case when the surrounding of a system is just seen as a second system. Besides by a thermal contact and heat exchange, work can be done on a system and end up as thermal energy by friction. Then, the surrounding introduces no entropy into the system. Instead, all energy exchanged as work and thus by a directed motion ends up in the system with disorder generated by friction at the contact surface, and thus increasing the entropy in the system.

What does this mean for energy, and especially how is thermal energy described in the concept of energy?

3.3.3.2 External energy without / with an effect from the surrounding

Looking at external energy, the same is true again for thermal energy as before for deformation, excitation, ionization chemical, and nuclear energy: we choose the system as all particles involved in the change. For thermal energy, as for deformation energy, this means many particles in bulk matter, not just few like in a change of composition. The system of particles can have kinetic energy as a whole, or as a whole be in a gravitation, electric, or a magnetic field. As long as these refer to the whole system, and do not affect what we focus on, they are already treated previously in section 3.1. Consequently, there is nothing new to discuss here about external energy. Thermal energy is related to systems at rest, and solely internal energy.

3.3.3.3 Internal energy without / with an effect from the surrounding

We will now see how to describe thermal energy, corresponding to $p \cdot dV$. For deformation energy, the pressure is the generalized force, as we know. If two systems interact at a common surface, they are in equilibrium if the pressure is the same in both systems; then, their volumes remain unchanged and consequently there is no energy change associated with deformation. The same way, if two systems interact at a common surface, they are in equilibrium if the temperature is the same in both systems. This expresses the **0th law of thermodynamics**, which states that **"if system A is in thermal equilibrium with system B, and system B is in thermal equilibrium with system C, then A is also in thermal equilibrium with C"**. Replacing one system with a thermometer it becomes the above statement **"two systems are in thermal equilibrium if both have the same temperature"**. Consequently the **temperature** T is the generalized force in thermal energy. It is commonly given in **Kelvin**, K, or degree **Celsius**, °C, related by

$$T / K = T / °C + 273.15 .$$

Eq. 3-47

And the generalized displacement? If two systems interact at a common surface, they are in equilibrium if the temperature is the same in both systems; then, there is no heat transfer, thus that their entropies remain unchanged. Comparison with deformation shows that entropy corresponds to volume. Thus, **thermal energy** is energy associated with a change of the entropy, and its change dE_{th} is described, using macroscopic values, by

$$dE_{th} = T \cdot dS .$$

Eq. 3-48

If a system exchanges energy as **heat**, then

$$dE_{int} = dE_{th} = T \cdot dS = dQ .$$

Eq. 3-49

But disorder, and with it thermal energy, can increase even if no disorder comes from outside by heat transfer, due to work and friction at the surface or by internal conversion. This is expressed by the **Clausius inequality**

$$T \cdot dS \geq \delta Q \quad \text{or reorganized} \quad dS \geq \frac{\delta Q}{T} .$$

Eq. 3-50

3.3.3.4 Energy conservation and work - energy theorem

Like for deformation, excitation, ionization, chemical, and nuclear energy, when dealing with thermal energy we do, for convenience and lack of detailed information, usually look at values observed macroscopically. Then, we can only observe that energy is conserved, but we cannot understand it because we cannot see its origin. However, since we know that materials are composed of particles we can understand it. From the previous discussion, we know that the internal energy of all atoms, molecules etc. in materials is the sum of their energies in the fundamental energy contributions, meaning kinetic energy and potential energy due to the 4 fundamental interactions. If energy is conserved in a single interaction it is also conserved in many. Quantum mechanics only affects the energy differences that are allowed. So, it doesn't matter if quantum mechanics plays a role or not. It is obvious that energy is conserved without an effect from the surrounding by a force, pressure, or any similar. If there is such an effect, including heat transfer or friction, energy is conserved as soon as the surrounding is included in the energy balance.

Expansion of the work - energy theorem by heat

The discussion of the "exchange" of energy initially started with just work and kinetic energy in section 2.1.3.4, being combined in the work - kinetic energy theorem. Then, in section 2.1.3.5, potential energy was integrated on the energy side, resulting in the work - energy theorem. We will now make another expansion to include heat.

If friction happens inside a system, e.g. if large convection currents become smaller and end up as thermal motion (Fig. 3-46), energy of energy contributions / forms with order inside the system is converted to thermal energy; thus there is no exchange of energy. If friction happens at a boundary of a system, e.g. when two pieces of wood are rubbed past each other, energy of energy contributions / forms with order outside the system is converted to thermal energy. In this case, work is done on the system, such that energy from the surrounding ends up as thermal energy in the system (Fig. 3-45). So far nothing has to be changed. When thermal energy is however also the source, that means when a system is in thermal contact with another system with thermal energy (Fig. 3-43), e.g. just its surrounding, then the situation is different. At the boundary, the thermal contact, collisions between atoms and molecules lead to an exchange of energy such that the thermal energy on one side increases while it decreases on the other. How does this fit into the work - energy theorem? There is no directed force acting for a distance from the surrounding, and therefore there is no work done on the system. The answer was introducing the term "heat". While work is energy exchange by a directed force that acts a distance, heat is energy exchange by undirected, random forces on a microscopic level, associated with disorder. Thus, the expansion of the work - energy theorem to include heat is

$$\Delta E = W + Q \qquad \qquad dE = dW + dQ$$
$$\qquad = \Delta E_{ext} + \Delta E_{int} \quad \text{or} \quad = dE_{ext} + dE_{int} \; .$$

Eq. 3-51

We could call it **work and heat - energy theorem**. It is important to note that because of its relation to microscopic effects, heat exchange only changes the internal energy. Thus, related to Eq. 3-51 is what is called the **1st law of thermodynamics**, which was stated in the 1880s by R. Clausius as **"the change of the internal energy of a system is equal to the amount of heat supplied to the system, and less the work done by the system"**. In its written form it is

$$\Delta E_{int} = -W + Q \quad \text{or} \quad dE_{int} = -dW + dQ \; .$$

Eq. 3-52

It is often said that the 1st law of thermodynamics is equal to energy conservation, but this is not correct. It is based on energy conservation, but its statement is about energy exchange between a system and its surrounding, and specifically the associated change of the internal energy of the system, and states that energy is exchanged as work or heat. Further on, in contrast to the work - energy theorem and the general statement in Eq. 3-51, the convention in Eq. 3-52 is that work is work done by the system on the surrounding, such that it has the opposite direction than heat supplied to the system; thus the "–" sign in Eq. 3-52. The reason for this is to relate directly to the technical application of thermal engines, which are supplied with heat and deliver work. The 1st law of thermodynamics is thus a statement to describe thermal engines, based on the observation that energy is conserved. The difference of Eq. 3-52 to Eq. 3-51 referring only to internal energy E_{int} is thus also not surprising as the treatment of thermal engines is usually with them being at rest and with no other interaction such that E_{ext} is unchanged; Eq. 3-52, focusing on E_{int}, is then a consequence of Eq. 3-51.

187

Differences, differentials, state and path functions

The notation in Eq. 3-51 and Eq. 3-52 looks somewhat strange, something we already realized when introducing the work - energy theorem in section 2.1.3.5, and similar as with momentum and impulse. This is just because it follows the statement that the change of the energy, a difference, is equal to the energy exchanged as work or heat, thus not a difference but an amount. As discussed in section 1.2, "Δ" commonly denotes a larger change e.g. between an initial and final point, while differential changes, which are infinitesimal small differences, are commonly denoted differently by "d". Then, when looking at small changes in general, the common notation is to write $dE = dQ \pm dW$, $W = \int dW$, $dW = F \cdot ds$ or $p \cdot dV$, and $Q = \int dQ$, $dQ = T \cdot dS$.

But this is not everything yet as we can see if we look at the meaning of the differentials and differences again and different situations they are used in. In general, the change of a value A along a path from an initial point i to a final point f is the sum of the changes by small steps dA, written as integral

$$\Delta_{i,f} A = \int_{i}^{f} dA \, .$$

Eq. 3-53

If $\Delta_{i,f}A$ is independent of the path, then dA is called an **exact differential**, denoted by d. If $\Delta_{i,f}A$ is path dependent it is called an **inexact differential**, and as additional information to specify the path is necessary, to signify the difference "δ" is used. For example, the height of the peak of a mountain above sea level is independent of the path how it is reached, but the length of the way to get to the peak depends on the path chosen. Properties that only depend on a state, and not on the path how it is reached, are called **function of state**, or **state function**, while properties that depend on the path how a state is reached are called **path function**, or **process function**.

The internal energy of a system in a certain state is independent of the path / process; it is thus a state function (like its volume or its temperature). But the energy exchanged as work and as heat is a path / process function, which is why cyclic heat engines can do work from heat supplied to them. Thus, changes of the internal energy are exact differentials, while work and heat are not. Consequently, the precise differential form of Eq. 3-52 is

$$dE_{int} = -\delta W + \delta Q \, .$$

Eq. 3-54

3.3.3.5 Energy changes, processes, and boundary conditions

Now, why are work and heat path or process functions? To find the answer we need to discuss some new aspects about energy forms of internal energy, things we avoided up to this point to make the discussion of the individual energy forms less complicated, and because the aspects are general and should thus be treated after discussing all energy forms in a final wrap up.

In a **thermal engine** a material, the **working medium**, undergoes changes. Its internal energy E_{int} is a function of the kinetic and potential energy of its particles, thus a function of state. The pressure p, volume V, temperature T, and entropy S, are also functions of state. After a full cycle the working medium is again in the same state as at the start (marked by a ○ at the integral)

$$\oint dE_{int} = -\oint \delta W + \oint \delta Q = 0 \,. \qquad \textbf{Eq. 3-55}$$

Thermal engines usually use a gas as working medium, placed in a cylinder with a piston to change the volume of the gas (Fig. 3-41), thus that work is done on the gas or by the gas. In a full cycle there is net work done by the engine. How is this possible? The work done in a closed cycle is not zero

$$W = \oint \delta W = \oint p \cdot dV \neq 0 \,, \qquad \textbf{Eq. 3-56}$$

and thus deformation energy $dE_{def} = p \cdot dV$ is also a path or process function and consequently an inexact differential; in fact we should thus write δE_{def}. The way this is possible is by varying p during the cycle; if p changes during the cycle then the thermal engine can do net work in a closed cycle. Specifically, p is smaller during compression and larger during expansion. How is that possible? It is by the gas having a different temperature. If the temperature of the gas is higher during expansion, then its pressure is higher; the higher temperature is achieved by previous heating. On compression the temperature must be lower by previous cooling. As energy is conserved, the heat supplied to the engine in a closed cycle is also not zero

$$Q = \oint \delta Q = \oint T \cdot dS \neq 0 \,. \qquad \textbf{Eq. 3-57}$$

Consequently, thermal energy $dE_{th} = T \cdot dS$ is also a path or process function and an inexact differential; in fact we should thus also write δE_{th}.

Is this something that holds only for thermal and deformation energy?

By using differential calculus we can get more insight. The entropy S and volume V are independent variables, such that we can write (Reif 1985)

$$dE_{int} = T \cdot dS - p \cdot dV$$

$$= \left(\frac{\partial E_{int}}{\partial S}\right)_V \cdot dS + \left(\frac{\partial E_{int}}{\partial V}\right)_S \cdot dV \; ; \qquad \textbf{Eq. 3-58}$$

the V and S right below the brackets signify that the value in the bracket is for constant V, respectively S. Eq. 3-58 has two consequences. The first is

$$T = \left(\frac{\partial E_{int}}{\partial S}\right)_V \quad \text{and} \quad -p = \left(\frac{\partial E_{int}}{\partial V}\right)_S , \qquad \textbf{Eq. 3-59}$$

thus defining T and p. By further differentiation follows the second one

$$\left(\frac{\partial T}{\partial V}\right)_S = -\left(\frac{\partial p}{\partial S}\right)_V . \qquad \textbf{Eq. 3-60}$$

It is one of several relations, called **Maxwell relations**, which relate various values of materials, here T, p, V, and S. While S and V are independent, all four are not. And if more energy forms are included in Eq. 3-58 by adding terms like $E \cdot dP$, $B \cdot dM$, and $\mu \cdot dN$, then E, B, and μ can be defined similar to Eq. 3-59, and then corresponding relations like Eq. 3-60 can be derived. What we found for thermal energy and deformation energy is thus general for energy forms of internal energy: their changes depend on boundary conditions. They are not state functions but instead path or process functions. Consequently, looking at a gas at given T and p we cannot say how much thermal energy it contains and how much deformation energy. Thus, a system does not "contain" or "have" internal energy in these energy forms. This is different for the fundamental energy contributions as their value can be calculated from information on all particles in a system. The energy in any of the energy forms of internal energy is the energy associated with a change of an observable variable, like S, V, P, M, N, during a process at conditions of e.g. constant volume etc. specifying T, p, E, B, μ. For example, deformation energy is the energy associated with a change of volume, during a process at conditions of e.g. constant entropy. Hence, the energy form is just "deformation energy"; it is only "elastic" at certain conditions.

3.3.3.6 Quality and usefulness

We know now that energy is conserved, that changes in the energy forms of internal energy, and with them work and heat, depend on conditions and thus the path of a process. And we also know that entropy tends to increase. These things affect energy conversion processes, so let's have a closer look.

In a **heat engine**, also called **thermal engine**, the exchange of work and heat with an amount of a medium, the system, changes its internal energy, specifically thermal and deformation energy. In a cyclic engine, the medium undergoes in a simplified way two processes, forward (1) and backward (2), such that (using U for E_{int} here, as is common notation in thermodynamics)

$$\Delta U_1 = Q_1 - W_1$$
$$\Delta U_2 = Q_2 - W_2$$
$$\Delta U_1 - \Delta U_2 = (Q_1 - W_1) - (Q_2 - W_2)$$
$$= (Q_1 - Q_2) - (W_1 - W_2) \, .$$

Eq. 3-61

In a cyclic engine, both processes lead back to the initial state of the medium when the cycle is closed; the change in internal energy is then 0, thus

$$0 = (Q_1 - Q_2) - (W_1 - W_2).$$

Eq. 3-62

W and Q however depend on the process and not on the state. Thus, if the two processes are identical no net heat or work was done during the cycle, while when they are different heat is "converted" to work or vice versa. This is the basis of all thermal / heat engines.

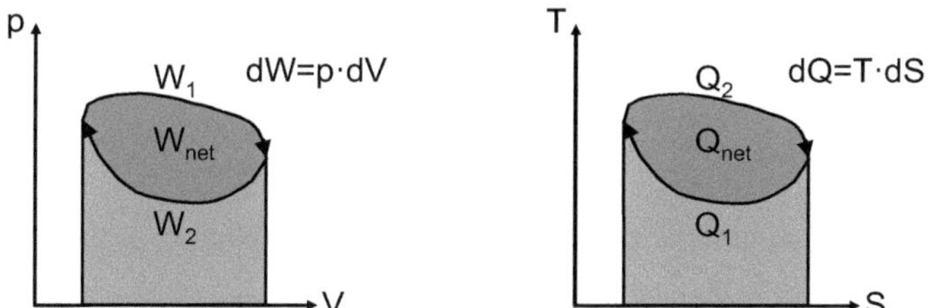

Fig. 3-47 Thermodynamic cycle in a p - V diagram and a T -S diagram

While the path dependence gives some freedom for conversion, generally limited by energy conservation, the increase of entropy limits the direction.

How much heat can be "converted" to work? Ideally, in a full cycle the net heat input is completely converted to work as output, so Eq. 3-62 becomes

$$W_{out} = Q_{in} - Q_{out}.$$

Eq. 3-63

This can be done in two ways: closed and open cycle

- closed: an amount of the medium is heated ($\rightarrow Q_{in}$), does work during expansion ($\rightarrow W_{out}$), is then cooled ($\rightarrow Q_{out}$) and the cycle is repeated

- open: the heated medium ($\rightarrow Q_{in}$) does work ($\rightarrow W_{out}$), and is then discarded ($\rightarrow Q_{out}$), and replaced by new heated medium to repeat the cycle

For the following calculation on energy and entropy there is no difference. Using Eq. 3-50, for a thermal / heat engine

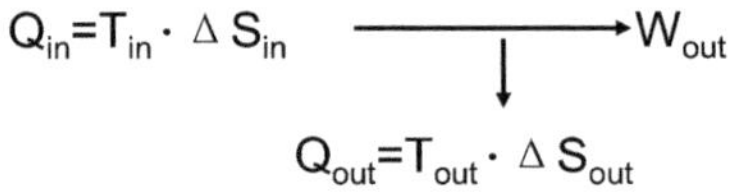

$$Q_{in}=T_{in} \cdot \Delta S_{in} \longrightarrow W_{out}$$
$$\downarrow$$
$$Q_{out}=T_{out} \cdot \Delta S_{out}$$

Fig. 3-48 Schematic thermal / heat engine with "losses"

This means that in a cyclic thermal / heat engine, to generate work W_{out} from heat Q_{in}, it is necessary to discard the entropy via Q_{out}. This is energetically a loss, but it is necessary because the entropy and thus disorder never decreases and has to go somewhere, and it can not leave with W_{out}.

In the ideal case (reversible process)

$$\Delta S = 0 = \Delta S_{in} - \Delta S_{out}.$$

Eq. 3-64

From Eq. 3-64 and Fig. 3-49 follows

$$\frac{\Delta Q_{out}}{\Delta Q_{in}} = \frac{T_{out}}{T_{in}}.$$

Eq. 3-65

From Eq. 3-63 we can then calculate the ideal **efficiency** of a heat engine as

$$\text{efficiency}_{he,ideal} \; \frac{\Delta W_{out}}{\Delta Q_{in}} = \frac{\Delta Q_{in} - \Delta Q_{out}}{\Delta Q_{in}}$$

Eq. 3-66

$$= 1 - \frac{\Delta Q_{out}}{\Delta Q_{in}} = 1 - \frac{T_{out}}{T_{in}}.$$

192

This is the best possible, thus ideal efficiency of a thermal / heat engine, valid for constant temperatures and no other losses than ΔQ_{out}; it is between 0 and 1 because $T_{out} < T_{in}$. The output of work (as a desired useful energy) is higher if the heat input is at a higher temperature. This is in accordance with Eq. 3-50, which shows that heat at high temperature carries less entropy / disorder, and is therefore more "similar" to work, which carries no entropy / disorder at all. While the conversion of heat to work is limited, work on the other hand can be converted to heat completely, for example by friction. We understand that from the discussion of the microscopic basis. Because work can be converted completely to work or heat, while for heat there are limits in the conversion to work, work is considered more useful than heat. Consequently, the **quality** of an energy is measured by how much work it can do: it is 100 % for any kind of work and less for heat according to Eq. 3-66. Heat at ambient temperature cannot be used to do any work as $T_{in} = T_{out}$; it is "useless", which is described by assigning a quality of 0 %. This shows that the definition of energy as "ability to do work" is wrong, as thermal energy delivering heat at ambient temperature cannot do any work.

3.3.3.7 Enthalpy, heat, and deformation work

There is a topic left, related to path dependence, which was mentioned already several times but not discussed yet: the **enthalpy** H; it is defined as

$$H = U + p \cdot V .$$
Eq. 3-67

The enthalpy is not path dependent, as U, p, and V are not path dependent. This is quite useful, e.g. in Hess's law (sections 3.2.4.4, 3.2.6.2, and 5.3.1). By differential analysis follows $dH = dU + d(p \cdot V) = dU + p \cdot dV + V \cdot dp$. So

$$dH = dU + p \cdot dV \quad \text{if } p = \text{const.}$$
Eq. 3-68

Thus, dH is the sum of dU and the expansion work done by a system on the surrounding at constant p. From Eq. 3-67, using $dU = dQ - p \cdot dV$, follows

$$dH = dQ + V \cdot dp .$$
Eq. 3-69

So dH is the sum of dQ and the work due to flow of a volume V against dp. Additionally, at p = const., enthalpy difference equals the heat exchanged. This is useful for calculating heat exchange e.g. in chemical reactions and many other applications. According to Çengel and Boles 2002, the term enthalpy derives from the Greek word "enthalpien", meaning "to heat").

193

3.3.4 Energy conservation and conversion – more examples

3.3.4.1 Determination of dS and dQ

Thermal energy is different in several ways compared to the other energy forms of internal energy. The change of the thermal energy $T \cdot dS$ is still the product of a generalized force and a generalized displacement, like $p \cdot dV$, but the entropy change dS is not related to a position, we also don't see it macroscopically, and we don't even have technical means for observation. The change of entropy dS is thus quite different from dV, dN, dM, or dP. Consequently, a change of internal energy by thermal energy cannot be calculated by simply multiplying the temperature T and the entropy change dS from individual measurements. How can we then practically deal with this?

A way to get $p \cdot dV$, without knowing p and dV, is by the related work dW. In the same way, we can determine $T \cdot dS$ from the heat exchanged dQ, if there is no other source, thus not external friction or internal conversion. And how can dQ be determined? Simply by generating the heat from work. There are three ways. The first way is by doing work in a way that is easily measured, and its full conversion to heat. Today this is commonly done by electric heating. In his experiments to determine the heat capacity of water, Joule used mechanical work that was converted to heat via friction; the work is easy to calculate from the applied force and the distance it acted. This leads to the second way. The ratio of the heat transferred to a body and the resulting temperature change is called heat capacity, with

$$dQ = C \cdot dT = m \cdot c \cdot dT \, . \qquad\qquad \textbf{Eq. 3-70}$$

Using the first option the **heat capacity** of a body, C, or the **specific heat capacity** of the material it consists of, c, can be determined (calibrated), and in a second step then used to measure an unknown amount of heat from a measured temperature change. The third way is similar, using the heat flow as a result of a temperature difference through a body acting as **thermal resistance** R, or the **thermal conductivity** λ of the material it consists of

$$dQ = 1/R \cdot \Delta T = A \cdot \lambda \cdot \Delta T / \Delta l \, . \qquad\qquad \textbf{Eq. 3-71}$$

We can then use Eq. 3-49, with dQ and T, to calculate the entropy change. However, what is not accessible is internal conversion to thermal energy.

3.3.4.2 Conversion between internal energies of deformation and thermal energy

Up to now we used the term "elastic" in connection with kinetic energy and collisions, and also with deformation. In general, "elastic" means "flexible". In the context of energy, **elastic** means that the energy of an energy contribution / form associated with a change is retrieved in the same energy contribution / form and by the same amount when a change is reversed again. For this it is essential that no lasting "losses" occur by conversion to another energy contribution / form. For example, if two elastic balls collide, the kinetic energy is temporarily converted to deformation energy, but as it is elastic, it is fully retrieved. In an inelastic collision permanent losses occur.

Focus on deformation energy

Elastic deformation means that the deformation energy associated with a change of form is retrieved when the form changes back. For example, when a body is compressed the deformation work $dW = -p \cdot dV$ is done on it. It is retrieved if on the reverse deformation process p varies the same way. This is e.g. the case if there is no heat exchange; the system is then called **adiabatic** and $dE_{int} = dW$. If heat is lost it must be supplied back when the process is reversed, the same way as it was lost before. For example, when a gas is compressed, work is done on it, its internal energy rises, and also its pressure. If the gas is contained adiabatically, the reverse happens on expansion at identical conditions. The case is quite different if the gas expands into an evacuated vessel, thus without a pressure, called **free expansion**. This process is irreversible, and no work is done such that the internal energy is unchanged. The same effects are observed on solids and liquids, but the temperature changes are usually small.

Focus on thermal energy

What about thermal energy? Elastic means that the thermal energy associated with a change of entropy is retrieved when the entropy changes back. For example, when a body is heated, the internal energy changes by thermal energy $dQ = T \cdot dS$. It is retrieved if on the reverse process T is the same. This is the case if there is no work, meaning that the volume is unchanged; the process is then called **isochoric**. If work is done on the system the change of thermal energy can be increased, which is e.g. used in mechanical heat pumps. In thermal engines, work is done by the system.

3.3.4.3 Conversion between thermal energy or deformation energy to other forms of internal energy

Thermal energy and energy associated with a change of composition

Excitation and ionization, as well as chemical and nuclear reactions, are tightly connected with changes of thermal energy by conversion processes. The conversion of these energy forms that are set free on a microscopic level is hard to use for a macroscopic, directed action and thus for work. Common is that the energy is set free as kinetic energy of the particles involved in the process of the change of composition, and in bulk materials the corresponding energy is readily converted to thermal energy by collisions with the particles in the bulk material. This is what happens in a reactor for nuclear fission or fusion, and in chemical reactions like combustion.

In many processes thermal energy is also used for excitation, ionization, chemical reactions, and in fusion even for nuclear reactions. An example is the reaction of two hydrogen atoms to a hydrogen molecule or the reverse (Fig. 3-32). If a hydrogen molecule is rotating and vibrating the bond between the two hydrogen atoms is stretched, and thus less extra energy by other means is needed to destroy the bond to separate the atoms completely. Molecular rotation and vibration are parts of thermal motion, while the energy needed to destroy a bond is the basis of chemical energy. This shows how the temperature affects chemical reaction energies, and that in the extreme case thermal energy can be used to drive a chemical reaction.

Deformation in an electric field – piezoelectric effect

If an object made of a material with bound charges is exposed to an electric field, as shown in Fig. 3-16, the charges move somewhat. As a side effect, the dimension of the object can change such that the object it is deformed. Fig. 3-17 shows this with gravitation pulling the charges back, while in materials it is the inter-particle bonds. The effect, called **piezoelectric effect**, describes the use of electric work for deformation. By the same conversion, the deformation of a solid can be connected to electric work when the charges are separated by a mechanical force and thus create an additional electric field. The effect is e.g. used in loudspeakers and the reverse effect in microphones.

Deformation and bond energy between particles

Let's now think of stretching a solid with charges further, like an ionic crystal in section 3.1.7.1. As Fig. 3-24 shows, the force from the surrounding stretches the crystal, thereby increasing its length. The discussion is with regard to a single fundamental energy contribution; the ions have no kinetic energy, and potential energy only as electric energy due to the electric interaction. The stretching is initially fully reversible, thus elastic deformation. When rupture occurs it becomes irreversible, thus **inelastic deformation**. Part of the work is used for rupture, thus supplies the energy to destroy the bonds at the rupture interface and is thus irreversibly converted at rupture to another energy contribution / form. The energetic effort from the surrounding for deformation is partly stored as deformation energy, and partly lost.

Another form of inelastic deformation is **plastic deformation**; here, particles change position on a microscopic scale in a way that upon release of the force they do not fully go back to their original position. If ongoing, this is also called plastic flow; in contrast to ordinary flow, plastic flow occurs only at a significant force from the surrounding. Inelastic, plastic deformation is e.g. taking place in a car crash, often also combined with rupture.

3.3.4.4 Conversion between external energy and internal energy of deformation or thermal energy

Falling object that comes to rest

By including thermal energy, we are now able to write down a complete energy balance for an object that falls and finally comes to rest on the ground. Because we do not know the process path, we can discuss deformation and thermal energy only as changes, and thus use changes in the energy balance

$$
\begin{aligned}
dE_{tot} &= 0 \\
&= dE_g + dE_{kin} + dE_{def} + dE_{th} \qquad\qquad \text{Eq. 3-72}\\
&= \quad dE_{ext} \quad + \quad dE_{int} \; .
\end{aligned}
$$

The total energy comprises initially external energy of gravitation, which is during the fall converted to kinetic energy, and then during impact on the ground to internal energy as deformation and "invisible" thermal energy. Now the initial, macroscopic energy that we can see does not "disappear" anymore; it ends up as internal energy of deformation and thermal energy.

For example, a piece of iron of 1kg falling 100m changes its gravitational energy by $m \cdot g \cdot h = 1kg \cdot 10N/kg \cdot 100m = 1000N \cdot m = 1000J$. In the process, its maximum velocity is by $m \cdot g \cdot h = \frac{1}{2} \cdot m \cdot v^2$ equal to about 45m/s. If it finally ends as thermal energy of the iron only, the resulting temperature rise is by $dQ/dT = m \cdot c = 1kg \cdot 449J/(kg \cdot K)$ about $\Delta T = 1000J/449J \cdot K = 2.2K$. For water, with $c = 4187J/kg$, the result is only $\Delta T = 0.24K$. This is why the conversion to thermal energy is often not observed without technical means.

Moving object on a flat surface that comes to rest

In section 2.1.1.1 we saw that if no force acts on an object, the object will stay at rest if it is at rest, or keep moving at the same velocity and direction. This is expressed by Newton's 1st law. And we already discussed that common observation is that a force acts, the force of friction, which makes a moving object slow down and finally come to rest. We also understand it now from the concept of energy: kinetic energy is converted to thermal energy by friction at the contact surface. With a rolling ball the effect is hard to observe, but with a flat stone on a slope as shown in Fig. 3-49 it is easy.

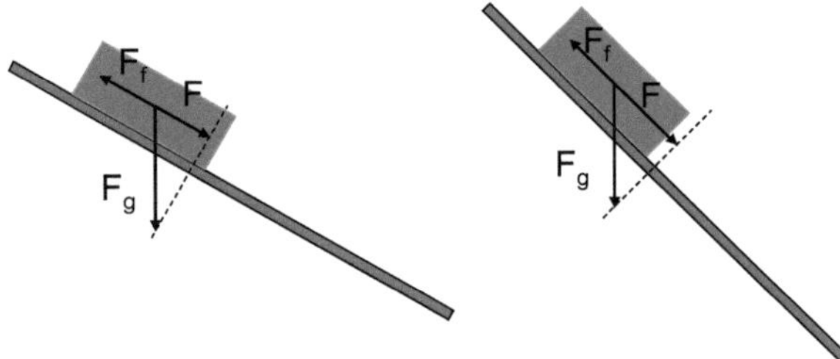

Fig. 3-49 Flat stone on a slope; left, the stone is at rest due to friction, right, the stone moves as friction is too weak

If only the force of gravitation F_g would act, the stone would slide, as its component F drags it down the slope. Due to the force of friction F_f at the contact area between stone and slope the net force can however be zero; then the stone stays at rest. If the inclination increases, F increases while F_f doesn't, such that finally a net force results and the stone starts to move. Careful measurements show that F_f is different if the stone moves or not. What is different? When moving, the electromagnetic interaction between the particles at the contact area between stone and slope leads to an exchange of energy (Fig. 3-45), while at rest it just causes a force.

3.3.5 Summary and conclusions

Section 3.3 started with an introduction to materials, including their classification as solid, liquid, or gas, and characteristic differences and properties.

Then, the set of energy forms of deformation energy and thermal energy was discussed. Deformation energy and thermal energy are both related to the relative motion and position of the particles in a material or a larger body made of a material, and thus a many-particle system.

Deformation energy is energy associated with a change of the form of a macroscopic body due to a directed force or a change of a directed force, e.g. $dE_{def} = p \cdot dV$. Microscopically deformation energy is due to a change of the relative motion and position of the particles, and thus a change of the kinetic energy and potential energy due to particle interactions. For an ideal gas, where there are no interactions between the gas particles except elastic collisions, the potential energy is zero, so there is only kinetic energy.

Thermal energy is energy associated with a change of entropy, and its change is described by $dE_{th} = T \cdot dS$. Microscopically, thermal energy is also associated with a change of the relative motion and position of the particles, related to a change of the kinetic energy and potential energy due to particle interactions. For an ideal gas, there is only kinetic energy, the same way as for deformation energy. However, for thermal energy the change is always by a change of entropy and thus disorder, while for deformation energy it is always by a force that is directed, thus with order (without disorder).

Thermal energy is therefore complementary to the energy of deformation; while deformation leads to an ordered change of the relative motion and position of the particles in the body, thermal energy is related to disorder. It is important to note that up to this point all effects were associated with order. Thermal energy is the only energy form that is connected with the disordered, often called random motion and position. As such, it is only possible in many-particle systems.

A change of thermal energy (change of kinetic and potential energy at a microscopic level) can be caused by collisions with particles of the surrounding at the surface, which themselves have a random motion. As the effect is not by a directed force or field anymore, a new term was introduced: heat. Therefore, now a system can exchange energy as work and as heat.

199

In any body or material, being solid, liquid, or a gas, there is internal energy by the motion of the particles, thus kinetic energy, and by their position due to mutual interactions, thus potential energy. And in many-particle systems the motion is random, thus without a preferred direction. In an ideal gas however, which has no interaction between particles except collisions, only kinetic energy of translation exists as internal energy. Thus, in an ideal gas deformation as well as thermal energy are based on the same random translational motion of the particles.

In all cases discussed before, every energy contribution / form had a distinct underlying effect as basis, and there had to be a connecting mechanism for conversion between them, for example particles having charge and mass. This is not the same for deformation and thermal energy; both energy forms are based partly on the same effect: the motion of the particles and their interparticle interaction. We can however clearly distinguish them by the associated order / disorder, e.g. if energy is exchanged with the surrounding, and this is the reason why both were introduced as separate energy forms. Deformation energy is related to an energy exchange by a directed force acting a distance at the system boundary, called work, and thermal energy is related to an energy exchange by undirected forces, called heat.

In a certain state of an ideal gas, e.g. described by p, V, T, it is impossible to say exactly how much of the internal energy or which part came from work by deformation and how much came by heat; it is all just kinetic energy. But there is a way to split them if we keep boundary conditions constant, e.g. in small changes. If the volume of the system is kept constant then there is no work and thus no deformation; any change of the internal energy can then be associated to thermal energy, the same way as in a solid; hence the use of "heat capacity at constant volume". If there is no heat transfer any change in the internal energy can be associated to work and deformation, the same way as in a solid; hence "adiabatic compression work".

For thermal energy there is no static case, but a static case can serve as a good starting point to understand what thermal energy is. Let's start with a solid (Fig. 3-38) at the lowest possible energy. This means the atoms and molecules are not moving, and located at a constant distance determined by the forces between them and their shape (this is not exactly possible according to quantum mechanics, but the difference is not important for the discussion here). At a higher energy the particles move; they have kinetic energy of vibration and for the latter some potential energy, like a pendulum. Due to collisions, the movement of the many particles relative to each other is disordered. When the energy is high enough, the motion will dominate directed interparticle forces such that particles can slide past each other; this is macroscopically observed as the material becoming a liquid. At even higher energy the particles can completely overcome the interparticle forces and move free, which is macroscopically observed as the material becoming a gas. If during these processes the form is constant, then no work was done and all energy is thermal energy due to heat transfer. If the volume is not restricted, the motion of the particles will tend to increase their bond length, such that expansion occurs in all dimensions. This effect, called thermal expansion, is then connected with work as it is a change of volume against the ambient pressure.

Let us now finish this chapter by looking back at all of what we discussed. Chapter 3 was generally dealing with complex situations with many objects. In its main sections, we looked at three sets of energy contributions / forms. In chapter 2, before, and as a basis, we discussed simple situations, which is 1 object in motion or in interaction with a 2nd object, including the fundamental energy contributions. They are only a function of the motion and position and result in absolute energy values. We saw that it is natural to take the zero point of energy of motion where the velocity of an object is zero, and the zero point of energy of interaction when the interaction vanishes, meaning when the distance between the two interacting objects is infinite.

In section 3.1 we then discussed many objects, but still having only a single fundamental energy contribution. Consequently, there are absolute values of energy. But already section 3.1.3.4 showed for the case of polarization of a system with many objects by an electric field from outside that it is hard to calculate the absolute value of the internal energy; it requires to deal with detailed information on the motion and position of all objects of the system. But its change is easy to calculate, by energy exchanged with the surrounding as work done on or by the system.

In section 3.2 we then discussed energy related to a change in composition. A change in composition by definition refers to an energy difference. Nevertheless, the energy of a defined composition has an absolute value, however it is rather hard to calculate because changes in composition refer always to microscopic particles that we cannot observe in sufficient detail.

In section 3.3, discussing deformation energy and thermal energy, we found another reason that makes it even more hard to calculate absolute values: both are based on the kinetic and potential energy contributions of the particles, and cannot be separated in all cases. The most prominent case is the ideal gas, but the effect of thermal expansion shows the general connection.

What is the consequence?

For a simple situation, which is 1 object in motion or in interaction with a 2nd object, it is possible to write down the external and internal energy as an absolute value, and this is what is commonly done. We can do the same thing for complex situations with many objects, as long as we look at the system as a whole, thus its external energy, as it is again a simple situation. This changes when looking at complex situations and their internal energy. When dealing with forms of internal energy it is hard and impracticable, often even impossible to deal with absolute values as the necessary detailed information is missing. In addition, energy forms related to what is going on in the material can affect each other, for example, when a material becomes liquid due to an increase in its thermal energy, this also affects the result when changing its deformation energy. Thus, small changes keeping other parameters constant are treated. While for forms of external energy absolute values of energy are given, for forms of internal energy usually only changes, specifically differential changes denoted by d are treated.

The internal energy of macroscopic systems, which are composed of many microscopic particles, is the field of study of thermodynamics. Its laws are

0th law of thermodynamics: two systems are in thermal equilibrium if both have the same temperature T

1st law of thermodynamics: the change of the internal energy of a system is equal to the amount of heat supplied to the system, and less the work done by the system, written as $dE_{int} = dQ - dW$ (Eq. 3-52)

2nd law of thermodynamics: the entropy in an isolated system can increase but never decreases, written as $dS \geq 0$ (Eq. 3-45)

The 1st and 2nd law give crucial limits for energy conversion: the 1st law is a consequence of energy conservation and thus limits conversion by amount. The 2nd law, basis for the Clausius inequality, written $T \cdot dS \geq \delta Q$ (Eq. 3-50), limits its direction. At the end, all energy becomes thermal energy.

4 Answer to what energy is

The analysis of simple and complex situations, done in chapters 2 and 3, covering a single to many objects, and a single to several fundamental interactions, is now finished. It is thus time to summarize the discussed things, analyze the findings, and see what we learned. We will do this in two steps.

In the first step, chapter 4, we look at the key issues, like what are energy contributions / forms, how is energy converted, and why is it conserved, to answer what energy is. In the second step, in chapter 5, we will then look at the whole concept of energy, specifically focusing on the different ways to describe the energy of systems by the different energy contributions / forms, and those to describe energy exchange between systems, e.g. work or heat. And we will compare the findings from the systematic analysis here with what has been developed from observations throughout history.

To start, we repeat again the scope of this book: the goal, and the crucial parts of the key issues, the stage, and the key ideas, which were introduced in section 1.1.

Goal

The goal of this book is to give a systematic, clear, and comprehensive treatment of the basics of energy: the concept and the physical basis.

Key issues

Energy is related to changes.

Energy appears in different forms. What are the energy forms and what is their basis?

Energy can be converted between the energy forms. How does energy conversion work? What are the mechanisms?

Energy conversion is limited, first because **the total amount of energy is conserved**, and second because **entropy tends to increase. Why is energy conserved? What is entropy, and why does it increase?**

All combined will explain the concept of energy and its physical basis, thus answer the question "what is energy?"

Stage

All of us have some common, basic concepts about the world around us. **Space refers to where things are, and time is used to describe when, specifically to describe processes of change. Things we are familiar with are objects, like a stone, a person, the moon ... Matter refers to what the things are made of; specifically large objects consist of smaller ones, commonly called particles. Forces relate to why and how things change, specifically the position of objects (including small particles) by motion (change of position). The forces are due to interactions between objects.** Thus, these concepts form the stage of knowledge in most parts of the book. In addition to these concepts, science started to describe things by laws. **Laws just state what is observed reliably, usually in many experiments under controlled conditions. The laws then allow everybody to make predictions, which is why they are so useful.** Common experience is about macroscopic objects that have a mass. The laws describing their motion and interaction using the concept of force are called the laws of **classical mechanics**, and Newton's laws of motion focus on their basis.

Key ideas

The **key ideas**, used to start the discussion in this book, are briefly:

1. Macroscopic observations of energy conservation or conversion on systems with many objects / particles, like materials, are understood from their microscopic origin.

2. For a comprehensive treatment, microscopic models and macroscopic observations must be discussed together in a single text, and extended beyond just internal energy in materials.

3. The origin of energy conservation and conversion is in simple situations, which is 1 object in motion, and cases of interaction with a 2nd object. Complex situations are just a combination of simple situations.

4. To be able to "understand" things it is necessary to stay within the validity of classical mechanics, specifically Newton's laws, as long as possible; this also means including microscopic situations.

5. In addition, a discussion of the historic background regarding general ideas and use of words is also crucial in an introductory text.

4.1 Summary and analysis of limits

4.1.1 Summary of the previous discussion

We experience our world as being composed of objects that interact with each other, and thereby affect each other. The objects can be large (Fig. 1-1) and thus visible, like a stone, a planet or the sun, or be small and invisible, like atoms and molecules in materials, their nuclei and electrons. We cannot see directly what goes on on the microscopic scale, but knowing that there are microscopic particles we can imagine the microscopic world as a small version of the macroscopic world that we know from common experience.

The goal of this book is to give a systematic, clear, and comprehensive treatment of the concept of energy and its physical basis. For the treatment to be comprehensive we need to cover all scales of magnitude. And we need to do it in a systematic and clear way. As the physical basis does not justify a clear boundary between microscopic and macroscopic scale or situations, any choice would be somewhat arbitrary. The discussion was thus not guided by the size. Instead, we have chosen to discuss things by their complexity. Because complex situations are composed of simple situations, as shown in Fig. 4-1, the basis of the common principles are simple situations.

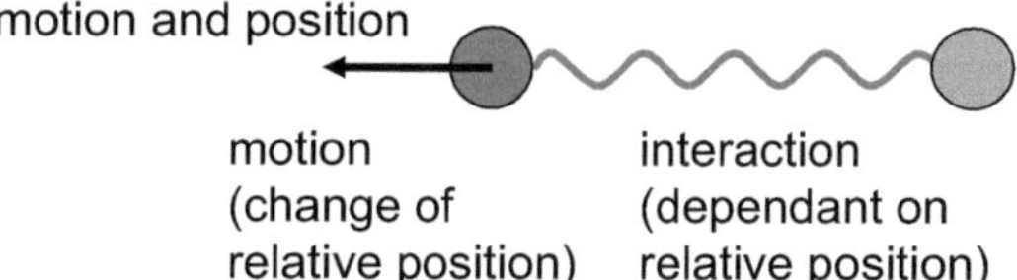

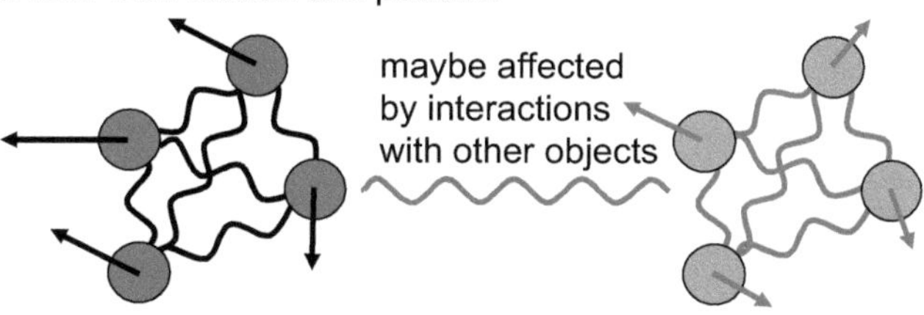

Fig. 4-1 Simple situation (top) and complex situation being composed of many simple situations (bottom)

A simple situation is the interaction between 2 objects and how the interaction affects them. The interaction between 2 objects is always by one or several of the fundamental interactions, which are gravitation, electromagnetic, and the two nuclear interactions. As gravitation acts on objects of common scale, usually with no significant effect by the other fundamental interactions, it was the choice to start the systematic discussion in chapter 2.

In section 2.1 "Basic terms, concepts, and laws" we thus started by looking at 1 object without or with interaction with a 2nd object by the fundamental interaction of gravitation, sometimes just represented by the corresponding force, or by collision where the interaction is unspecified and just called "direct contact". Situations with 1 or 2 objects are observed easily and frequent in common situations, and were the start of the development of the different concepts in history. To keep things simple, the discussion focused on linear motion and one type of force, the force of gravitation.

In the course of the discussion, the most important energy related terms and concepts, specifically force and Newton's laws of motion related to forces, momentum and its conservation, energy and energy contributions / forms, conversion between them, and overall energy conservation were discussed.

What are energy contributions / forms? An energy contribution is anything that contributes to the total energy; "contribution" has a general meaning. The second term, energy form, is more restrictive. An energy form refers to contributions that are connected to direct observation. The different energy forms have come up during the investigation of macroscopic effects. The energy forms introduced were kinetic energy for motion, and for position, related to a second object interacting by gravitation, gravitational energy.

How does energy conversion between energy contributions / forms work? For conversion a connection between them is necessary. For example, an objects mass connects its kinetic energy with its gravitational energy.

Why is energy conserved? For simple situations, this is revealed quite easy. If there is no interaction nothing changes, especially the motion of an object does not change. Thus, while motion and the associated energy is visible, and therefore commonly in the focus, the real basis of energy are the invisible interactions and their associated potential energy. From looking at the physical basis, kinetic energy is just stored energy from an interaction. As Newton's 3rd law states, forces in an interaction are always pairwise, having the same magnitude but opposite sign. And we saw that the same holds for changes of momentum and changes of energy. Actually, the concepts of force, momentum, and energy all describe interactions, and the real physical basis is that an interaction between two objects acts on both objects the same way but in opposite directions. Thus changes of momentum as well as changes of energy cancel out. This mathematically allows to say that energy and momentum are conserved. Conservation is thus often interpreted as indicating that something exists that has a constant amount, but the real basis is just that changes are pairwise and cancel out. For forces it is easy to understand, because forces and Newton's 3rd law are a common experience. For momentum it is also not hard to understand as momentum and its conservation in a collision are observable. For energy it is however very hard to understand from common experience because there are different energy contributions / forms and because we only see changes associated with them e.g. of the velocity for kinetic energy or of the height for gravitational energy (Fig. 4-2). But this is also what makes the concept of energy so useful.

$$\begin{array}{lcl} \text{total energy} = & \text{energy contribution} \quad + & \text{energy contribution} \quad + \;\; ... \\ (\text{``conserved''}) & \text{or form} & \text{or form} \\ \text{const.} = & & \end{array}$$

$$0 = \quad ... \cdot \boxed{\text{observed change}} + \quad ... \cdot \boxed{\text{observed change}} + \quad ...$$

$$\begin{array}{lll} \text{free fall:} & E_{kin} = 1/2 \cdot m \cdot \boxed{v^2} & E_g = m \cdot g \cdot \boxed{h} \\ \text{thermal expansion:} & dE_{th} = T \cdot \boxed{dS} & dE_{def} = p \cdot \boxed{dV} \end{array}$$

Fig. 4-2 Schematic use of the concept of energy, relating changes

Regarding its use, **energy is a concept relating changes** (not their cause). While energy conversion connects changes associated with the different energy contributions / forms (Fig. 4-2), energy conservation limits them by their absolute value. E.g. for free fall it connects the changes of v and h.

While the interactions between two objects always act on both of them, there are different reasons for looking only at one object / side sometimes. The description of the energetic effect of an interaction can be done by the work - energy theorem. Work was introduced as the product of a directed force and the distance it acts on an object, resulting in a change of the kinetic energy of the object.

In section 2.2, still staying with simple situations with only 1 or 2 objects, two limitations were removed. First, objects are not only point-like but now also can have an extension; thus, besides location and linear translation, now orientation and rotation can play a role. And besides gravitation the other fundamental interactions, which are electromagnetic and nuclear interactions, were discussed and the concept of a force field was introduced. These expand the work - energy theorem on the energy side.

The 4 fundamental interactions are the basis of all interactions that can occur, within the atomic nuclei, atoms, molecules, stars, the solar system ..., hence the term "fundamental energy contributions". No further interactions and thus forces are necessary for a comprehensive, detailed description. So, regarding its physical basis **energy is a concept to describe the effect of interactions between objects on the objects, specifically on their motion and position.**

However, to describe simple situations like the swing of a pendulum or the collision of two balls, the concept of energy is not necessary; the formulas are still simple using the concept of forces acting on the individual objects. It is thus common practice to describe things using the concept of forces without talking about energy. Energy and its conservation became useful when applied to complex situations where the application of forces is not practical or not possible, for example situations with many objects, with objects that are not seen such that the observation of changes does not relate to an individual object, when several interactions act combined, or when the type of interaction is unknown. In such complex situations we usually try to avoid the details, or we don't even know them. For example, the thermal energy of a stone related to the kinetic and potential energy of the atoms it is composed of. Complex situations are where the concept of energy and its conservation is not only useful, but often even the only practical option.

In chapter 3 we then discussed complex situations, which are composed of many simple situations, as shown in Fig. 4-1. As energy is conserved in a single interaction it must also be conserved in many interactions, no matter if they are in series or in parallel. And this also does not change if different interactions are involved; the fundamental interactions do not affect each other, this is why they are called fundamental. Their related forces and energies can thus be added. And the same way there is also nothing fundamentally new about energy conversion.

The discussion of complex situations with many objects in chapter 3 focuses on two general situations, which continue the previous discussion of 1 object in motion, and cases of interaction with a 2nd object. As Fig. 4-1 shows, the single objects are now replaced by many objects, which together we call a system. The first general situation is many objects, the system, without interaction with its surrounding, and the second is with an interaction with its surrounding, thus another system. As before, the interaction can be described without the 2nd part, by a force field as "origin" of a force, by an energy potential field for energy, or by work to describe the energetic effect of a force acting a distance.

Replacing the single object by a system of objects (Fig. 4-1) adds two new aspects: the system has energy due to the interaction between its objects and due to their motion, and this energy can also be affected by an interaction with its surrounding, thus actually another system. The new part, the energy associated with what goes on in the system of many objects, was given the name internal energy in the historic development of the concept of energy. Internal energy refers to the energy contributions / forms due to whatever goes on in the system. In contrast, energy associated with the whole system, not based on changes in it, should be consistently called external energy. Another possible option would be to call it merely non-internal energy. Regarding external energy the whole system is like a single object and can be described by the ways that were discussed in section 2.2: by its motion, and thus kinetic energy, or by its interaction with the surrounding, and thus potential energy. For the internal energy, a detailed description is complex due to the many objects and interactions. Often detailed information is not even available. It is thus common to describe just changes of the internal energy, which are due to energy conservation in the system with its surroundings equal to the work done on or by the system.

In section 3.1 the discussion of the cases in section 2.2 was continued for a system of many objects. We discussed the cases of kinetic energy and those with a single fundamental interaction, for point-like objects as well as now for objects with an extension. External energy can be due to the mass of the system, its net electric charge, electric or magnetic dipole moment, and also nuclear effects which are however only relevant on the scale of a nucleus. The discussion also showed that for gravitation there is no change of the internal energy due to a gravitation field from the surrounding if it is homogeneous, for electric energy it is $E \cdot dP$, and for magnetic energy it is $B \cdot dM$.

Discussing energy contributions regarding motion or just to a single fundamental interaction individually is the way to understand what energy is, why it is conserved, and how it is converted. At the end of section 3.1 a crucial point in the systematic discussion is reached. The discussion in section 3.1 was the continuation of the discussion of the fundamental energy contributions discussed section 2.2 now for many objects; consequently, now all possible effects are described for a single object as well as for many objects. The energy contributions are well defined; no part can be missing, and no parts might be counted twice. This seems to be the perfect situation to use the concept of energy, but it is often not. Why? Energy and its conservation connect changes (Fig. 4-2) e.g. of velocity and height if a pendulum swings. But it is only useful if the changes are expressed in a way that they refer to something we can observe. It is thus not surprising that the historic discovery of energy contributions followed the way of observable energy contributions, which are what is commonly called energy form. The start was kinetic and gravitational energy because they are already observable on the macroscopic level by the eye. But what is the effect that releases energy during a chemical reaction like combustion? Where does the energy go when an apple falls from a tree and finally lies on the ground at rest? These questions introduced to two new sets of energy forms that were discussed in the remaining sections of chapter 3, and are different from kinetic energy or potential energy related to a single fundamental interaction. Nevertheless, their basis is still the kinetic and potential energy from fundamental interactions. Historically, new energy forms like chemical energy and thermal energy were introduced without any initial understanding of their physical basis. But due to energy conservation they were still successfully integrated into the concept of energy, specifically energy balance calculations using values that describe macroscopic observations. Thus, initially energy conservation was just what was observed in experiments. The discovery of the microscopic physical basis, which is microscopic particles and their interactions, was much later as it required the discovery of the particles and a new way of describing the energetic interaction: quantum mechanics.

In section 3.2, the first set of such energy forms was discussed, comprising excitation, ionization, chemical, and nuclear energy. They are related to a change in the composition of particles, by the parts or by their arrangement: for excitation and ionization regarding the electrons in an atom or molecule, in chemical reactions regarding atoms and molecules in compounds, and for nuclear reactions within the nuclei of atoms. Therefore, the section started with an introduction to atoms and quantum mechanics. The related energies combine contributions of kinetic energy and potential energy from several fundamental interactions. Their practical description by what can be observed macroscopically is as a reaction energy, meaning as an energy difference between an initial and final state. Thus, only the amount of substance that reacts must be known; and this is what can be observed easily.

In section 3.3 the second set of such energy forms was discussed: deformation energy and thermal energy. Both are related to the kinetic and potential energy of all particles of a material or of a larger body made of a material, and therefore a many-particle system. For this, section 3.3 started with an introduction to materials. Deformation energy refers to the change of kinetic and potential energy when the form, meaning the size or the shape changes. The absolute value of the energy is in this case practically never calculated; a typical zero point is however without an external force, or just atmospheric pressure. A typical description of the energy change due to an external effect, by values that can be observed macroscopically, is for length change by F·dl directly, or for volume change by p·dV. Thermal energy also refers to the kinetic and potential energy of the particles, but not for a change of size and shape. Thermal energy is complementary to the energy of deformation; while deformation leads to an ordered change of the motion or position of the particles in a body, in the case of thermal energy it is with disorder, measured by the entropy S. The description of the energy change is by T·dS. It is important to note that up to this point all effects were associated with order. Thermal energy is the only energy form that is connected with a change of disorder in the random motion and position. As such it is obviously only possible in many-particle systems. A change of thermal energy of a system by collisions with external particles at the boundary, which themselves have a random motion, is called heat Q, in contrast to work W which is by a directed force F acting a distance ds. A system exchanges energy by work and heat, as stated by the 1st law of thermodynamics. Overall disorder increases, as stated by the 2nd law of thermodynamics; thus, thermal energy can also increase within a system without exchange of heat by conversion from other energy contributions / forms to thermal energy. Therefore, entropy limits energy conversion by its direction, in contrast to energy conservation which limits energy conversion by the amount. For changes of thermal energy and deformation energy the latter shows Fig. 4-2.

4.1.2 Analysis of limits

Before we try to give a final comprehensive definition of energy, we need to discuss possible consequences of the limits that we set by the chosen stage. The chosen stage is the existence of objects (if small called particles) in space and time, which interact and thereby affect each other. In the course of the discussion (sections 2.1.6, 2.2.6, and 4.1.1) we realized that regarding its physical basis **energy is a concept to describe the effect of interactions between objects on the objects, specifically their motion and position.** The concepts of force and of momentum do this also (section 2.1.7). They all describe the effect of interactions between objects on the objects motion and position. And all of them have specific limitations and also advantages. The concept of forces is closest to our common experience, and therefore was chosen as the approach to discuss what happens in different situations, and to understand "energy". The basics describe Newton's laws of motion. What we need to do now is to analyze the stage and its effect on the result.

Limitations

The use of Newton's laws of motion is challenged on a microscopic scale because they fail to describe things like the behavior of electrons in atoms. Instead of what is called Newton mechanics, quantum mechanics is used. Then, what is different, and what isn't? First, by introducing the wave nature of objects their motion is not anymore described by a position or a path. And the energy in bound states becomes quantized. But quantum mechanics uses the same functions describing the interactions as using Newton's laws. Thus, quantum mechanics does not affect energy itself, but only which energy levels are possible. This becomes relevant if the particle wavelength is in the order of the dimension of the available space in microscopic situations. If not, the quantization of energy is too small to be observed, and the energy is correctly derived using Newton's laws. It is thus not surprising that in macroscopic systems quantum mechanics fully agrees with Newton mechanics and also that conservation of energy is still observed. And that energy is conserved in macroscopic situations requires that it is also conserved in microscopic situations. The meaning of energy therefore does not change, nor how energy is converted; conversion is still between the different energy contributions / forms that are connected via a particle having e.g. mass and electric charge. However, if there is no path or position we cannot use Newton's 3rd law to explain energy conservation.

Somewhat more critical is the interaction between atoms by electromagnetic fields and waves, as the challenge on Newton's laws of motion goes further. While we can understand the effect of an electromagnetic field in a macroscopic situation by its electromagnetic force, governed by Newton's laws, we cannot do the same on the microscopic scale for interactions with atoms. For this the concept of exchange particles is used, e.g. photons for the electromagnetic interaction. Photons describe the interaction within the concept of energy, including quantization of energy. Because photons "carry" energy, they are part of the energy balance to have energy conserved any time. However, photons have no mass and no electric charge, so they do not describe how energy is carried, or exchanged with an atom. Thus, photons are no particles like electrons; they are just a concept to describe quantization. But they are a concept that is still well within the limits of our imagination.

More critical is the existence of electromagnetic waves, and that they carry energy, without any objects involved. Even in empty space, an electric field can induce a magnetic field and reverse, without any charge being involved, and both fields have energy. This is far beyond our stage and imagination, where we have objects, which are affected by interactions between them. However, it is the objects that are the cause of interactions between them, and the interactions then in turn affect them. They both go hand in hand. Electromagnetic induction shows that this separation is however not needed. And the wave - particle dualism is another hint. But it is not the last one.

In nuclear reactions mass can be lost and become energy (section 3.2.5.4). And particles can even disappear completely, like in an electron-positron annihilation (reaction of matter with anti-matter). Then, the main foundation of our stage, that objects cause interactions, e.g. by their mass or charge, and that the interactions affect the objects, is lost. The fields representing an interaction do not only have energy, mass and energy are not even conserved individually. As Einstein showed, $E = m \cdot c^2$, so both are conserved together called conservation of mass-energy. Thus, the stage is left in such reactions, however the concept of energy is already modified to take the new effect into account. The situation is like with energy conservation a century ago: we observe mass-energy conservation, use it for calculations, but cannot explain and understand it as the common concept of objects that interact and thereby affect each other is not applicable. But such situations are rare, and do not occur in daily life at all. We do not have to bother about this limitation except in atomic physics.

Consequences for energy conversion and conservation

The stage set for the discussion was the existence of objects (if small called particles) in space and time, which interact and thereby affect each other, described by forces governed by Newton's laws of motion.

This stage does not only reflect common thinking about the world around us; as we saw, it also covers science and engineering, with few exceptions. Thus, the derivation of energy conservation and the ways of conversion from Newton's laws is widely valid regarding the type of interaction and size of objects. But it is limited to objects having mass, and beyond the validity of Newton's laws the derivation does not hold. But within our stage of objects in space and time, which interact and thereby affect each other, described by forces governed by Newton's laws of motion, derivations hold not only, we also can really understand things. As the systematic analysis showed, with the exceptions mentioned above, it is all about objects (if small called particles), forces between them, Newton's laws, and potential and kinetic energy. From them follow all energy forms, conversion processes, and energy conservation. If we would see all things in detail directly, there would be no "energy form". The energy forms simply originate from effects, which are related to energy and energy changes, and are classified according to macroscopic appearance and use: m and v, p and V, T and S, E and P, M and B, or μ and N for the composition of materials. For example, thermal energy is on a microscopic level related to changes of the kinetic and potential energy of invisible particles. After the systematic analysis, we probably should better ask ourselves "Why should energy not be conserved?" Energy conservation rather seems to be a logic consequence! Every interaction is symmetric to the interacting objects, which is expressed in the concept of forces by Newton's 3rd law, and for energy and momentum leads to pairwise changes with same magnitude and opposite sign. Therefore, it is more precise to use an adaptation of Newton's 3rd law for energy, like "In general, to every action associated with a change of energy there is always an equal and opposite reaction (action and reaction energy), such that energy is overall conserved." Energy conservation only seems to contradict everyday experience because of two misunderstandings. First, we often take the energy form that we can see as the (total) energy, and think it is lost when converted to an energy form that we can not see any more. And second, we say "consumed" about the initial energy form when using fuels, but it is just that energy form which gets less and not the total energy.

4.2 Conclusions – what is energy?

We have now summarized the systematic discussion in chapters 2 and 3, and we have analyzed it with respect to the stage and the limits. As a result, we can conclude that the basics of energy can be understood looking at the motion and interaction of objects, from simple to complex situations.

What's next? We should now give a clear and comprehensive answer to what energy is. But what is a good answer? And what should it contain? The information in an answer, or a definition, is given for others to use in some context, such that it is commonly necessary to customize the information in content and wording to a specific group of people with a certain knowledge and purpose. Thus, the answer should differ between groups.

The answer, or a definition, should include (see the goal and the key issues)

- the physical basis, meaning what energy is, its nature, and

- the concept, meaning what energy does, its use, boundaries, so mainly:

> energy is related to changes,

> energy appears in different contributions / forms,

> energy can be converted between the energy contribution / forms,

> energy conversion is limited, because energy is conserved and because entropy tends to increase

Let's start with a very short summary of our findings regarding these points.

Things we are familiar with are objects, like a stone, water that flows etc. Matter refers to what things are made of; specifically larger objects consist of smaller ones, called particles. Everything that happens is due to the motion and interaction of objects. Interactions are at the basis of what happens.

In simple situations, which is 1 object in motion or in interaction with a 2nd object by one of the fundamental interactions, things are easy to understand. Regarding the physical basis, energy is a concept to describe the effect of interactions between objects on the objects motion and position. Optionally, other concepts are force and momentum, all having specific characteristics.

In complex situations, which is many objects in motion and interacting, we can only understand the basic effects due to the complexity. In materials, where the objects are small particles, we also cannot see the detailed motion or interaction. What we are commonly interested in is changes we observe macroscopically, like a change of volume, or of an amount of a substance. These changes are still based on simple situations, but in most cases by complex combinations. Thus, for quantitative answers we need calculations.

Regarding its use, energy is a concept describing the effect of interactions between objects on the objects motion and position in a way that even in complex situations energy is connecting the resulting macroscopic changes. Fig. 4-2 shows this, for a simple situation, and also for a complex situation. The terms related to an interaction are a product of a force and a displacement, in complex situations called generalized forces and displacements. The observable changes resulting from an interaction are quantitatively related to changes in energy contributions; if the changes are macroscopically observable they are called energy forms. For example, the kinetic energy of an object is related to its velocity v, the gravitational energy to its height h, changes of thermal energy to changes of entropy S, and changes of deformation energy to a change of shape, for example by a change of volume V.

In an interaction, the changes that are resulting from the interaction, and with them the changes in their energy contribution / form, change together; this is commonly called energy conversion. For example if an object falls, gravitational energy is converted to kinetic energy, thus relating a change of height to a change of velocity. And the changes are related quantitatively.

Interactions act on the interacting objects in a symmetric way; as a result, changes of the involved energies cancel out, also if many particles interact. If the energy in one energy contribution / form increases, it decreases by the same amount in others. Mathematically, the total value of energy is then constant, which is commonly called conservation of energy. In the concept of forces the symmetry of interactions is expressed by Newton's 3rd law, in the concept of momentum it results in the conservation of momentum. The different options show again that energy is no "thing", but just a concept. Instead of just saying that energy is conserved, we better say, or remember, that every interaction comprises an action and a reaction, associated with an action and reaction energy of equal magnitude and of opposite sign.

Energy conservation limits energy conversion by the amount; in addition, the tendency of entropy to increase limits energy conversion by direction. In time, thermal energy increases at the expense of other energy forms.

Comparison with definitions found in literature

Before trying to give an own answer on what energy is, let's have a look at the information given in common literature and analyze it. While doing this, we should remember again that the information in a definition is given for others to use, such that it is commonly necessary to customize the given information in content and wording to the specific group of people addressed.

The following definitions of "energy" can be found in an English dictionary (Hornby 1983): a) force, vigor; capacity to do things and get things done b) powers available for working, or as used in working c) (science) capacity for, power of, doing work: electrical kinetic potential … This definition generally mixes different terms, as in colloquial use. Regarding science, it says about energy that its real nature is being a "capacity", limited to work.

In sci.-tech. textbooks it is not uncommon to just use the term "energy" without giving a definition or explanation, e.g. Reif 1985, and Çengel 1998. Others refer only to energy within a system, therefore to internal energy, like Cleveland and Morris 2009, Brown et al. 2015, Çengel and Boles 2002, and Sears and Zemansky 2016, or they refer again only to energy exchange between systems as work and heat, like Atkins 1990, Cleveland and Morris 2009, and Brown et al. 2015. None of them explains what energy really is. Actually, if we combine the definitions "energy is the ability to do work" and "work is energy exchanged by a system …" we get something like e.g. "energy is the ability for energy exchange by a system …"; meaningless …

The following is a review, comparison, and discussion of definitions that are relevant regarding the question of what energy really is. A few of them, collected by David Watson, are listed on his own website (Watson 2014). He surveyed 15 textbooks on thermodynamics, heat transfer, and physics. Only those definitions useful to explain what energy is are now discussed. A definition that he collected, quoted in a book called "Energies" by Vaclav Smil (Smil 1999), attributed to David Rose, is that energy "is an abstract concept invented by physical scientists in the nineteenth century to describe quantitatively a wide variety of natural phenomena." This is very close to the result here. Another definition he collected, from the book "Thermodynamics", written by Virgil Moring Faires and Clifford Max Simmang (Faires and Simmang 1978), a college textbook, is "Energy is inherent in all matter. Energy is something that appears in many different forms which are related to each other by the fact that conversion can be made from one form of energy to another. Although no simple definition can be given to the general term energy, E, except that it is the capacity to produce an effect, the various forms in which it appears can be defined with precision." Dave Watson's own definition is "Energy is a property or characteristic (or trait or aspect?) of matter that makes things happen, or, in the case of stored or potential energy, has the "potential" to make things happen. By "happen", we mean to make things move or change condition. Examples of changes in condition are changes in shape, volume, and chemical composition (results of a chemical reaction). There are also changes in pressure, temperature, and density which we call a "change of state" in thermodynamics. Phase changes, such as changing from solid to liquid, or liquid to vapor, or back the other way, are also good examples of condition changes. Something happened! Without energy, nothing would ever change, nothing would ever happen. You might say energy is the ultimate agent of change, the mother of all change agents. Whenever anything happens or changes there is an energy change." Also pointing to changes are the definitions in Mehling 2016 "... energy is the reason why things change ...", in Çengel and Boles 2002 "energy can be viewed as the ability to cause changes", and "energy is the ability of a system to cause exterior impacts, for instance a force across a distance" by Quaschning 2005. A definition by Cleveland and Morris 2009 for "energy" (physics) is "the use of this capacity to perform useful functions for humans, such as heating or cooling buildings and enclosures, powering vehicles and machinery, lighting, cooking foods and so on." This one is more a definition for colloquial use, not having scientific parts.

Walker et al. 2014 state "Energy is a number that we associate with a system of one or more objects." They continue "If a force changes one of the objects by, say, making it move, then the energy number changes. After countless experiments, scientists and engineers realized that if the scheme by which we assign energy numbers is planned carefully, the numbers can be used to predict the outcomes of experiments and, even more important, to build machines, such as flying machines. This success is based on a wonderful property of our universe: energy can be transformed from one type to another and transferred from one object to another, but the total amount is always the same (energy is conserved). No exception to this principle of energy conservation has ever been found." This is close to the result derived here, however missing the common origin of conservation and conversion in the individual interactions.

Summarizing the content of the collected definitions roughly gives

- Energy allows to describe qualitatively a wide range of phenomena, or energy allows to produce an effect / to do things / get things done / to perform useful functions / that makes (has the potential to make) things happen; to make things move or change in conditions like shape ... why things change

- Energy comes in different forms, it can be converted between the different forms, and overall energy is conserved (no explanation is given)

- Energy is an abstract concept, number, measure, or it is a property / characteristic, or it is a capacity / ability, reason / cause, released / available / needed to ... associated with objects, inherent in matter.

We can now discuss these different approaches.

Beginning with the first part, the use of energy to describe qualitatively a wide range of phenomena, is certainly correct, but not specific. The same applies to energy to produce an effect / to do things / get things done / to perform useful functions / that makes (has the potential to make) things happen. These are good for colloquial use. For a sci.-tech. use, energy to make things move or change in conditions like shape is quite specific.

The second part is undisputable; this is common knowledge in textbooks.

The critical part is the last part that tries to describe the nature of energy. Energy is a concept, like force and momentum, as the discussion showed. Force and momentum describe things close to everyday experience, consequently the use of the terms in colloquial as well as sci.-tech. use is similar. Energy in contrast is not directly experienced, e.g. we see the velocity of an object, but not its kinetic energy, and we see the height of an object, but not its gravitational energy. This has led to a blurred description in many ways. Energy describes the effect of interactions between objects on the objects and thereby describes how observed changes are interrelated quantitatively. In energy balance equations, energy shows up as a number, a value associated with energy forms. So, is energy a capacity / ability, or reason / cause? We saw that changes that we observe are related to changes of the energy in an energy form, but the energy itself is not the reason / cause; it is the change of energy, more precise the reason / cause for changes is an imbalance in forces. And they are in turn related to changes of the energy forms.

And energy is also no capacity / ability. An ability is e.g. being able to solve mathematic problems. If we have money, we could say that it enables us to do, buy … or that we need money to …; but having money is not an ability. The same holds for energy. We need energy to do work, it enables to …, however in a sci.-tech. context this does not tell anything about its nature. Energy is also not only related to matter or objects; there is also the energy of electromagnetic fields, so this restriction must not be part of a definition. Energy should also not be called a property or characteristic, as this would imply something directly connected e.g. with an object or an amount of a material, but gravitational energy only exists in relation to the origin of the field of gravitation, or combustion energy only exists when oxygen is present and depends on pressure. The basis of energy are the interactions. However, if the conditions are clear, expressions like capacity / ability, or reason / cause allow a simple discussion of the changes. The price paid is that if somebody goes beyond a simple discussion, asking about the basics, misconceptions occur. Energy makes it possible to change things, so we could say energy "enables" to …. And terms like energy is released / available / needed can be necessary to discuss energy in a wider audience. Again, they are useful to make a discussion of energy simple, however the simplifications cause problems when thinking about what energy really is. For energy, a practical definition is often scientifically incomplete, partly incorrect, or misleading, and a scientifically correct one is usually not practical for many people. Thus, a definition as "energy enables / is needed to do work (mechanical, electrical …) and to transfer heat" misses an explanation, but serves the need of many people as it reflects what goes on in their daily life as "users" of energy.

Any explanation of what energy is, and thus ultimately a good definition, must comprise all these aspects: that it is not a substance or anything alike but a concept, that interactions between things are in the focus, especially that it relates changes in different things due to interactions between them, and how it does that. Because of the wide variety of knowledge and interest in the audience, it is necessary to give a definition for people in a sci.-tech. context and another definition for common people, reflecting the respective needs of the different groups.

Based on what was just said about energy, we can now improve the preliminary definition from section 2.1.6. We will do this as far as possible in a general manner, however based on the discussion within the chosen stage.

Sci.-tech. definition of "energy"

Revising the preliminary version, a **sci.-tech. definition of "energy"** is

Energy is a concept, introduced in science and also colloquially used, describing how many changes that we observe in things are correlated.

Energy connects very different changes we can observe, e.g. of the velocity of an object, its shape, an amount, in a simple, quantitative way. The observable changes are quantitatively related to an energy form, for example, the kinetic energy of an object is related to its velocity, deformation energy to a change of shape, or chemical energy to a change of an amount of substance. By an interaction, the energy in the involved energy forms changes, which is commonly called conversion of energy. As a direct consequence, the involved observable values also change. For example, if an object falls free, gravitational energy is converted to kinetic energy, thus relating a change of height to a change of velocity.

The different changes are interrelated quantitatively, as the sum of the energies in all energy forms is conserved, called energy conservation. This gives the impression of energy being some thing that is conserved. Why energy is conserved can be understood by rewriting the statement: changes of energy overall cancel out. This is just a direct consequence of interactions being symmetric on the objects or particles that interact.

Colloquial definition of "energy"

Generalizing the preliminary version, a **colloquial definition of "energy"** is

Energy is a concept, introduced in science and also colloquially used, describing how many changes that we observe in things are correlated.

Energy connects very different changes we can observe, e.g. the velocity of an object, its shape, even an amount of substance, in a simple way. This is done by so-called energy forms related to the observed changes, for example kinetic energy to velocity, deformation energy to shape, chemical energy to an amount of substance etc. Their changes are related, called a conversion between the energy forms, even quantitatively by overall conservation of energy.

When focusing only on one part of a conversion process, we can say that energy enables / is needed e.g. to do work and to transfer heat.

4.3 Summary

Despite its importance in science and engineering, definitions of energy given in literature are not satisfactory, especially regarding what energy is. Definitions usually state that energy has different contributions, that energy is conserved but without an explanation, that conversion between the different contributions is possible, and that energy can be used to do certain things. While this information is sufficient for many people as it describes well how energy is used and for what, it misses an explanation and a clear statement what energy is. An attempt was made here to close this gap.

The discussion showed that the basics of energy can be understood looking at the motion and interaction of objects, from simple to complex situations. We saw the origin of energy conservation and conversion is in simple situations, which is 1 object in motion, and cases of interaction with a 2nd object. Complex situations are a combination of simple situations. Macroscopic observations of energy conservation or conversion on systems with many objects / particles, like materials, are understood from the microscopic origin. To be able to "understand" things it was necessary to stay within the validity of classical mechanics, thus with forces and Newton's laws of motion. We saw that this limitation is not as severe as what it seems at a first glance; even when quantum mechanics must be applied and for energy of electromagnetic fields we can still understand why energy should be conserved. The discussion of the historic background regarding general ideas and use of words has showed where and why many misconceptions occur.

The new definition suggested here includes the explanation, within its limits, and clearly specifies how energy is used and why it is so important. The impressing success of the concept of energy comes from combining effects that are macroscopically so different and doing this without the need to go into any detail. Energy and its conservation relates the changes of so many effects to each other, understood and not understood, reversible and not reversible, mechanical, thermal, … and it allows not only to describe, but also to affect, and even control and actively manage changes. In that respect it is much wider applicable than momentum, which refers only to the product of force and applied time; momentum connects only motion, while energy connects also displacements and other things of an incredible variety and meaning for our life. But this is also what makes it hard to understand without a systematic analysis of the concept of energy, as we have done here.

5 Complete concept of energy

The systematic and detailed discussion in chapters 2 and 3, from simple to complex situations, revealed that the physical basis of energy is in simple situations; complex situations are just a combination of simple situations. Complex situations, with many objects and several fundamental energy contributions, are understood from simple situations of just 1 object in motion, and cases of interaction with a 2nd object. Thus, within the stage of the book, **energy is a concept to describe the effect of interactions between objects on the objects, e.g. their motion and position.** And knowing the basis of energy we could also check for limitations by the stage that we have chosen. In chapter 4 we reviewed and summarized the answers to the key issues, which are what energy is in general, what energy contributions / forms are, how conversion between them works, and why energy is overall conserved.

In chapters 2 and 3 we discussed things following complexity, by the fundamental energy contributions and energy forms in parallel to see the basis of the energy forms. But the practical use of the concept of energy is commonly by just using one: fundamental energy contributions or energy forms.

In chapter 5 we will therefore look at the concept of energy again, first just by the fundamental energy contributions, and finally just by energy forms. We will compare what we developed in chapters 2 and 3 to what has been developed throughout history and is used today in science and technology. Historically, the concept of energy developed stepwise; people discovered a new effect, and then modified the concept of energy to integrate it, e.g. by a new energy form. This way the view is piecewise, and so is understanding. The systematic discussion here gives a complete view on the different parts of the concept of energy, and also how these parts are related to each other. But there are still some remaining questions left that need to be answered. For an energy balance to be correct, knowing that energy is conserved is not enough; in addition, no energy must be missing or counted more than once. As the concept of energy historically developed from experience and not from systematic analysis, we should not just take that for granted. However, because energy balance calculations work, we can expect the energy forms to have a systematic structure, and we will try to discover it in this chapter. This will allow us to see the whole concept of energy, its difference to thermodynamics, and will help to check the definitions of the energy forms.

We start with the physical basis and general issues of the concept of energy, and then summarize the general "tool box" that we have to discuss energy. Afterwards we look at the concept by the fundamental energy contributions, and finally by the energy forms, thus the connection to common experience.

5.1 Physical basis and general issues

Objects, systems, interactions, and energy

The physical basis of energy is the interaction between objects, connected to their relative position and changes thereof by motion. This is most obvious and thus easy to understand by looking at only two objects (Fig. 5-1, top), the most simple situation. Things are harder to understand when many objects interact with each other, thus complex situations (Fig. 5-1, bottom).

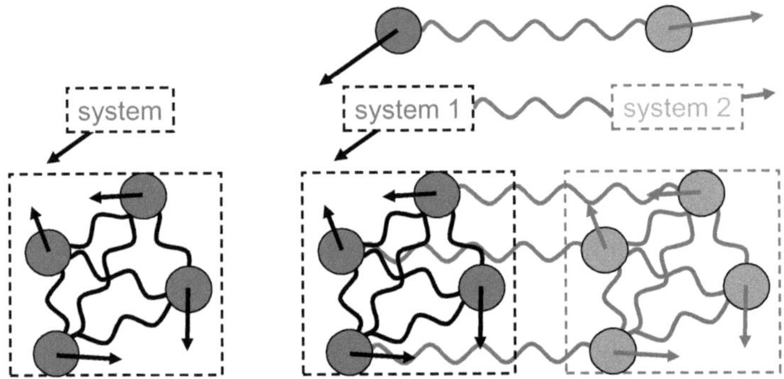

Fig. 5-1 Objects, interactions, simple and complex situations, systems

To discuss complex, but also simple situations, the term "system" is useful. A **system**, in general the sum of parts that interact with each other, is a quantity of matter, a number of objects, or a region of space, to be studied. What a system refers to is a question of choice with regard to matter, objects, or region of space in general, but also regarding the specific situation. In the discussion here, a system is in general a number of objects; it can be a single one or many of them (Fig. 5-1). The specific objects chosen depend on the situation investigated. For example, a number of objects can be considered a single system, but also be split in two parts with each of them treated as a system. The **boundary** of a system, a real or imaginary surface, separates a system from its **surrounding**, e.g. another system (Fig. 5-1).

Energy contributions / forms, and energy conversion

There are 4 **fundamental interactions**: gravitation, electromagnetic interaction, strong and weak nuclear interaction. The magnetic is just a relativistic effect of the electric interaction. The fundamental interactions are the basis of any interaction; any interaction is by one or by several of them combined. The physical basis of energy is the interaction between two objects, due to a property like mass or charge and the relative position of the objects, or due to the motion of an object, because of its mass and its (relative) velocity. Energy related to position is called **potential energy** E_{pot}, and energy related to motion is called **kinetic energy** E_{kin}. Due to the 4 fundamental interactions there are 5 **fundamental energy contributions**. They are the basis of all energy contributions / forms, by just one, or by several combined.

An **energy contribution** is anything that contributes to the total energy; "contribution" has a general meaning. The related term, energy form, is more restrictive. An **energy form** refers to changes that are macroscopic, thus directly observable, and are where energy was discovered in history. For example, kinetic energy is called an energy form when it refers to the energy of a car in motion, but kinetic energy of the atoms in a stone, which we can't see directly, is called an energy contribution to the thermal energy of the stone, an energy form that we can experience by its temperature. Important is that there are two terms with somewhat similar meaning, but used in somewhat different ways; no definition of them was found in literature. Due to interactions between objects the energy in the different energy contributions / forms can change. The observation that one increases while another one decreases is called **energy conversion** or **energy transformation**. In literature, energy contribution and energy form are often used synonymously, as well as conversion and transformation.

Overall energy conservation and entropy increase

Overall, **"energy cannot be created or destroyed; it can only be transformed from one form to another"**, as was stated by J.R. von Mayer, J.P. Joule, and H.L.F. von Helmholtz already in the years 1842-1847, based on observations. In other words, **energy conservation** and **energy conversion**.

Besides overall energy conservation, overall **entropy** increase limits the possibility and the degree of energy conversion.

5.2 The "tool box"

To make the concept of energy useful, it is necessary to transfer its physical basis and general issues into a tool box, with formulas expressing laws like energy conservation, and further classifications to structure the discussion. We will start discussing energy "of" systems, then energy "exchange between" systems, and finally energy conservation and conversion.

5.2.1 Energy of systems – internal and external

In simple situations of 1 object in motion or in interaction with a 2nd object there is energy associated with the motion of an object, thus kinetic energy, and energy associated with the relative position of both objects by their mutual interaction, thus potential energy. Looking at a system of several or many objects (Fig. 5-1, Fig. 5-2), instead just a single, adds another aspect. A system with more than 1 object has energy in it, again energy associated with the motion of the objects and energy associated with their position by mutual interaction. And such a system has still also energy as a whole due to its overall motion or due to its interaction with another system (Fig. 5-2).

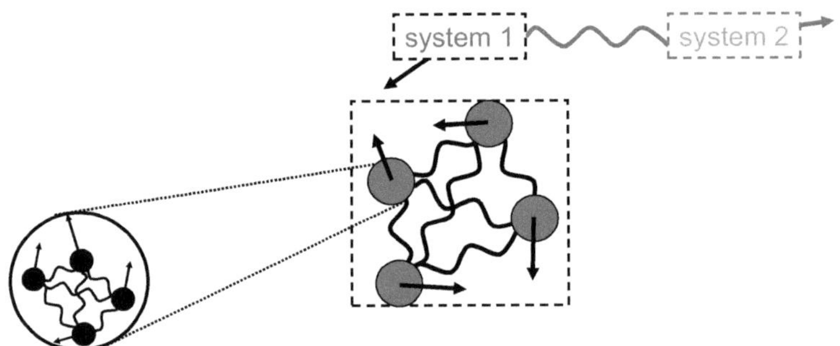

Fig. 5-2 System of objects having external energy as a whole (top) and internal energy (bottom), including objects having intrinsic parts

For example, a ball filled with air has energy associated with it as a whole: by the motion as a whole when flying it has kinetic energy, and depending on its position as a whole to the earth by gravitational interaction it also has gravitational energy. In addition it has energy in it, called internal energy, by the motion of the air molecules inside as kinetic energy.

Despite its importance, definitions of the terms in literature are imprecise. The following is a collection of definitions found for the "internal energy". Brown et al. 2015 state that "The internal energy E of a system is the sum of all kinetic and potential energies of the components of the system." And similarly, Sears and Zemansky 2016 "… define the internal energy of a system as the sum of the kinetic energies of all its constituent particles, plus the sum of all the potential energies of interaction among these particles." Both definitions state what belongs to internal energy, but not its nature, meaning what "not internal" is. The definition by Sears and Zemansky 2016 even refers to particles, thus limiting internal energy to just materials. The same holds for the following. Çengel and Boles 2002 define that "The sum of all the microscopic forms of energy is called the internal energy of a system and is denoted by U", not specifying what microscopic forms are. Significantly more helpful is the following. Cleveland and Morris 2009 define internal energy (thermodynamics) as "a property of a system that includes the kinetic energies of the individual particles of the system, the interaction energies between the particles, and the intrinsic energies of the individual particles, but that does not include the kinetic and potential energy of the system as a whole." It still refers to particles but clearly states that the definition is for the field of thermodynamics, and it adds specifically the option of having intrinsic energies of the particles (Fig. 5-2). More important, it states that there is also kinetic and potential energy of the whole system, and distinguishes it from internal energy, however without giving a name. Together, the definitions reflect the limited scope of thermodynamics, mainly looking at what goes on in materials, thus speaking about particles, their kinetic and potential energy in a system, but not of the whole system.

A complete concept of energy must not have these limitations. In general, we want to divide energy in energy related to what goes on in a system and energy of a system as a whole. Both parts can comprise kinetic and potential energies. As the term internal energy is already widely used for energy related to what goes on in a system, it is logic to use the term external energy for energy related to the system as a whole (Mehling 2016). Another option would be "non-internal" energy. And a general definition must also not be limited to materials, thus we should talk about objects instead of particles, or try to avoid them in general to be open to include also energy of fields. Mehling 2016 made a first attempt for a general definition; it is modified here to fit to the use of the terms object and system here.

"External energy is energy associated with effects of systems as a whole. Thus, external energy comprises the kinetic energy by the motion of a system as a whole, and the potential energy due to interaction of a system as a whole with another system by their relative position." This is like a simple situation. Examples are translational or rotational kinetic energy by the motion of a whole system, and potential energy due to the mass of a system by its interaction by gravitation with another system with mass. **"Internal energy is energy associated with effects within a system. Thus, internal energy comprises the kinetic energy of the parts of a system by their relative motion, in particular to the centre of mass of the system, and the potential energies due to interactions between the parts of a system by their relative position."** Examples are the potential energy e.g. by gravitation between the parts of a system due to their mass, by the electric interaction due to their electric charge, or by several fundamental interactions e.g. by a covalent bond between atoms. So, related to a change of internal energy are also changes of the composition of parts of a system.

Any energy of a system is either internal or external, thus that

$$E_{sys} = E_{int} + E_{ext} \quad \text{and} \quad dE_{sys} = dE_{int} + dE_{ext} \ . \qquad \text{Eq. 5-1}$$

Internal energy E_{int} and external energy E_{ext} are complementary, meaning nothing is missing or counted more than once, like kinetic and potential energy, or the fundamental energy contributions. This is crucial.

5.2.2 Energy exchange between systems

Interactions show up in the concept of energy in two notably different ways. For example, the earth and moon interact by gravitation. Together they have gravitational energy. In addition, (Fig. 3-9) the interaction of the moon with the earth by gravitation causes the tides and motion in the earth's interior, thus a change in the gravitational energy of parts of the system earth (then a rise in thermal energy of the earth), and a loss of kinetic energy of the moon. This result is commonly called an exchange of energy between both, moon and earth. When two systems interact, they can thus **exchange** energy

$$dE_{sys,1} \xrightarrow{\text{exchanged } E} dE_{sys,2} \ . \qquad \text{Eq. 5-2}$$

Specifically, they can exchange their "own" energy, comprising their internal energy as well as their external energy by motion (or interaction with a third system, not shown in Fig. 5-2), but not the interaction energy with the system they exchange energy as this one belongs to both systems together. As the example shows, the energy contributions / forms of both systems, which change, can have a different basis than the interaction for exchange. Again, we subdivide the energy exchanged, e.g. with regard to it carrying entropy or not as heat and work, sometimes even more detailed by the interaction e.g. as electric work. And for convenience or for a general view, often only one system is mentioned, then the second system is its surrounding.

5.2.3 Energy conservation and conversion

Experience has shown that the amount of energy in the universe is constant, which is called conservation of energy. Energy conservation was derived from experimental observations. We saw that, within the set stage, the basis of energy conservation is that in an interaction energy changes are pairwise, with equal magnitude but opposite sign, so that energy changes cancel out. Thus, for the **total energy** E_{tot} holds

$$dE_{tot} = 0 .$$
Eq. 5-3

In an **isolated system**, which is a system that has no mass and no energy exchange across its boundary, mass and energy are conserved. Thus for an isolated system (actually only if it is an inertial system) holds

$$dE_{sys} = 0 .$$
Eq. 5-4

The energy of a system can be subdivided in different ways, as we just saw, by energy contributions / forms (designated by n) thus that

$$dE_{sys} = \sum_n dE_n ,$$
Eq. 5-5

by external energy and internal energy thus that

$$dE_{sys} = dE_{ext} + dE_{int} ,$$
Eq. 5-6

or by both

$$dE_{sys} = \sum_n dE_{ext,n} + \sum_n dE_{int,n} .$$
Eq. 5-7

These equations relate the different energies of a system.

But they also relate changes of interacting systems if we see both systems interacting together. For energy "exchange" between systems, the energy of system (2) changes by the same amount as system (1) but with opposite sign, such that overall energy is conserved

$$dE_{sys,1} = -dE_{sys,2} \Leftrightarrow dE_{sys,1} + dE_{sys,2} = 0 \,. \qquad \textbf{Eq. 5-8}$$

As a consequence the different energy contributions / forms and the internal and external energies of both systems are also connected. Thus, in general, for energy conversion between different energy contributions / forms

$$\sum_n dE_n = 0 \quad . \qquad \textbf{Eq. 5-9}$$

Energy conversion between different energy contributions / forms is why energy is so useful. Energy conversion is limited by amount by overall energy conservation (and additionally by direction by overall entropy increase). A change of the energy in one energy form is always balanced by one or more other changes. Energy never disappears or comes out of nothing. What is meant when saying that energy is consumed, wasted, lost, or saved, is not energy in general but just a specific energy form of interest.

Ideally, for a correct energy balance calculation the energy is known as a set of energy contributions / forms that is complementary, meaning nothing is missing and nothing is counted more than once. The overall energy can be subdivided in different ways systematically, depending on a situation, what is interesting etc. The fundamental energy contributions are one option; as "fundamental" already says, they are complementary. But the energy forms are energy contributions that are simply connected to direct observation, just where energy was discovered throughout history. We thus cannot take for granted that the energy forms the way they were discovered and are used are complementary. And if they are, are they defined correctly? We already saw how different the definitions of "internal energy" are. Usually we just look at changes we know about, and assume anything else does not change. But of course we should try to do better. We will try to find out if the set of energy forms is complete, and if nothing is counted more than once.

But first we will look at the concept of energy using the fundamental energy contributions, thus kinetic energy and potential energy of the 4 fundamental interactions. It is a complementary set, and easier to understand.

5.3 Concept using fundamental energy contributions

The "size" of a system depends on the choice; it is not limited in any way. A system can be a stone (Fig. 5-3) that consists of millions of atoms, a molecule that consists of few atoms, an atom that consists of a nucleus and electrons, a nucleus that consists of nucleons, and even a nucleon that consists of quarks. A system is also the universe consisting of galaxies, the solar system, planets with moons, and also asteroids consisting of stones.

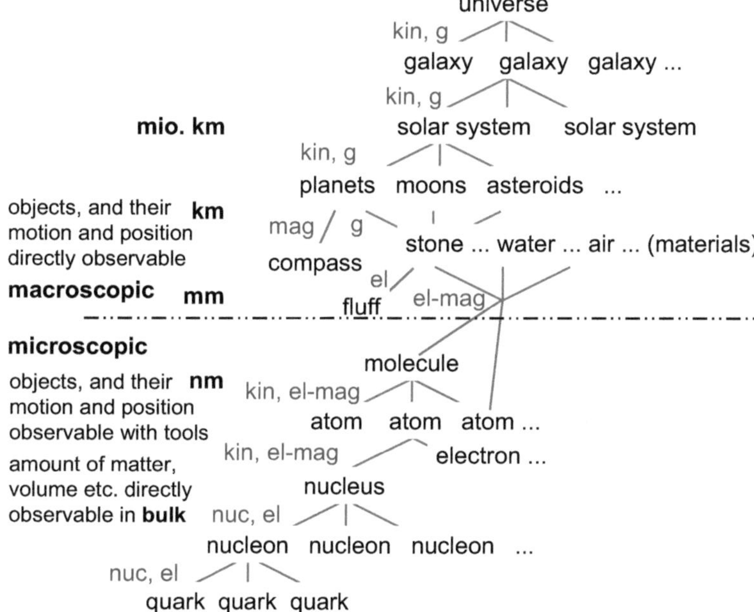

Fig. 5-3 Systems, objects, and fundamental energy contributions

In section 2.2 "Fundamental energy contributions" we discussed the 5 fundamental energy contributions, comprising kinetic energy and potential energy of the 4 fundamental interactions in detail, looking at simple situations. In section 3.1 "Energy related to a single fundamental energy contribution" we looked at systems, discussed their internal energy and external energy, and also discussed interactions between systems and energy exchange.

There were two reasons to choose this approach for the discussion here. First, discussing the fundamental energy contributions individually is the easiest way to understand what energy is, why it is conserved, and how it is converted. Second, the fundamental energy contributions cover everything; no part can be missing, and no parts might be counted more than once. At a first glance, this seems to be perfect to write down the energy balance of a situation and use the concept of energy for predictions, optimization etc.

However, there is a serious problem. The concept of energy must relate to observations that can be made more or less easily to be useful, but the fundamental energy contributions individually are not always suitable for this. There are several reasons. The first we already discussed in section 3.1: complexity. If a situation is too complex, a description by individual objects and energy contributions might be too complex to handle. Macroscopic objects are usually not in big numbers, but microscopic ones. Size is relevant in several ways, but is actually, with a single exception, not the real origin. Things smaller than about 0.1 mm can't be seen with the eye, meaning we cannot see objects with a smaller size, changes of their motion or position. It is thus not surprising that the historic development of the concept of energy started with easily observable things: kinetic energy and gravitational energy on a macroscopic scale (Fig. 5-3). But strictly, it is not a matter of the size, it is a matter of being able to observe the relevant information. Less common, but still observable on the macroscopic level, is the magnetic interaction acting on the compass needle, related to magnetic energy, and similar for the electric energy for fluff sticking to other objects. What is below the macroscopic level is not directly observable. For example, in a fire we see wood becoming ash and fumes rising, but not microscopic effects. Knowledge of the microscopic effects is also commonly not needed; even if we have detailed information, there is commonly no need or interest in it. We need to know how much wood has to be burnt to release enough energy, e.g. to raise the temperature of a desired amount of water by a certain value. The process is thus commonly not described by fundamental energy contributions but by the energy forms of chemical and thermal energy. Energy forms, when referring to internal energy, usually have a microscopic origin based on more than one fundamental energy contribution, and when quantum mechanics plays a role might not act individually, e.g. kinetic and electric energy of the electrons in atoms. In science we are interested in the details, have means to observe them, and computers for complex situations.

Let's start with fundamental energy contributions to be able to see later why and where energy forms differ, and if something is missing. For this, we look at energy of systems, being internal and external, at energy exchange between systems, at energy conservation and conversion in a system and in connection with energy exchange. We check if we can write down a correct energy balance, and if it is practical having changes that are observable.

5.3.1 Energy of systems – internal and external

The following repeats what we have discussed in sections 2.2 and 3.1 about energy of systems by fundamental energy contributions. The examples mentioned here, and many others, are shown in Fig. 5-3.

Kinetic energy

The basis of kinetic energy is the motion of a mass, which is described by its velocity. It has an absolute value, being zero at zero velocity.

As external energy, kinetic energy is associated with the motion of a whole system, specifically its center of mass, and described by its velocity. A system can have any size, thus be e.g. an electron, a stone, or a galaxy.

As internal energy, kinetic energy is associated with the motion of the individual masses within a system, described by their velocity. A system can be e.g. an atom comprising a nucleus and electrons, or a galaxy with millions of stars and planets moving around in it.

Gravitational energy

The basis of gravitational energy is the interaction of two masses by gravitation, affect by their distance. It has an absolute value, being zero at infinite distance. Because gravitation is comparatively weak, it is commonly only relevant if at least one of the masses that interact is huge.

As external energy, gravitational energy is associated with the interaction by gravitation between the mass of a whole system and that of another system, affected by their distance, e.g. a stone that interacts with the earth.

As internal energy, gravitational energy is associated with the interaction by gravitation between the individual masses within a system, affected by their distance. A system is e.g. all the parts of the earth, or the solar system with its planets, an interstellar dust cloud, or a galaxy.

Electric energy

The basis of electric energy is the electrostatic interaction between two electric charges, affected by their distance; electric dipoles are just a combination of two charges. It has an absolute value, being zero at infinite distance. Because the net el. charge of macroscopic bodies is usually close to zero, electric energy is usually only relevant in systems of microscopic size.

As external energy, electric energy is associated with the electric interaction between the electric charge of a whole system and that of another system, affected by their distance, and for dipole moments also their orientation. It is rare in macroscopic systems, except e.g. fluff sticking to clothes.

As internal energy, electric energy is associated with the electric interaction between the individual el. charges of a system, affected by their distance, for dipoles respectively their dipole moment also orientation. A system is e.g. a salt consisting of ions, or a metal consisting of ions and electron gas.

Magnetic energy

Magnetism is a relativistic effect of the electrostatic interaction, not an independent fundamental interaction. But it is practical to treat it separately. The basis of magnetic energy is the magnetic interaction between two magnetic dipoles, affected by their distance and orientation. It has an absolute value, being zero at infinite distance. It is relevant in microscopic systems, and in macroscopic systems e.g. in artificial magnetic fields.

As external energy, magnetic energy is associated with the magnetic interaction between the magnetic dipole moment of a whole system and that of another system, affected by their distance and orientation. A system is e.g. a compass needle in the magnetic field of the earth.

As internal energy, magnetic energy is associated with the magnetic interaction between the magnetic dipole moments of a system, affected by their distance and orientation. A system is e.g. the atoms in magnetized iron.

"Nuclear energy"

The basis of nuclear energy is the strong nuclear interaction between two color charges, affected by their distance, or the weak nuclear interaction e.g. in β-decay. Both interactions are relevant only at distances smaller than the diameter of nuclei, and commonly treated as internal energy of nuclei.

Discussing the energy of systems by the fundamental energy contributions relates directly to the fundamental interactions (or the corresponding forces) between the individual objects by their relative position, and in addition their relative motion for kinetic energy. It is the easiest way to understand what energy is, specifically the energy of systems. The fundamental energy contributions all have absolute values, by normalizing kinetic energy to zero at zero velocity, and by normalizing interaction energies (potential energies) to zero at infinite distance where the interaction vanishes. In addition, they describe internal and external energy on all scales of size in the same way. And there is another crucial advantage regarding the calculation of energies. Because the fundamental energy contributions are "fundamental", no energy can be missing, and none can be counted more than once. These are significant advantages using fundamental energy contributions.

Disadvantages show up if a situation is complex, having many objects and several fundamental interactions, if we do not have detailed information on the position and motion of the objects, or if we are not interested in details. As Fig. 5-3 shows, the fundamental energy contributions are useful when referring to macroscopic objects, where we can see the motion of an object or its relative distance to another object with the eye, where the situation is simple, and where we are interested in the details. This means for kinetic and gravitational energy, and few cases of electric and magnetic energy. Here, the fundamental energy contributions can also be called energy forms. What goes on at a microscopic level in detail is invisible, usually complex, and often not of common interest. Thus, common experience has led here to a different description by energy forms like thermal and chemical energy. But in science and technology we are often interested in the details, have options to see them, and tools like computers to deal with complexity. There, the fundamental energy contributions are often the prime choice. However, even in science and technology there are important exceptions. When a system is so small that quantum mechanical effects dominate, it can be hard to describe things by individual fundamental energy contributions. In an atom or molecule, the kinetic and electric energy of the electrons do not show up individually; instead, their sum is quantized, and is affected by interaction with the surrounding e.g. when absorbing a photon. Nuclear energy, when taken as the energy due to the strong nuclear interaction, is even limited to processes within nuclei, and never shows up alone. This makes it hard for scientists to discover, investigate, and determine its properties.

For these reasons, scientists developed tools to simplify calculations, and avoid details if not of current interest. Based on energy conservation, they allow a simple calculation of energy differences, but not absolute values. The first step is to tabulate the energy differences of common processes, e.g. per mass or amount of substance that is changing during the process. For more, energy conservation has two very useful, general consequences: (1) the energy difference of a process and of its reverse process have same magnitude and opposite sign, and (2) the overall energy change in a process is the sum of the energy changes of the process steps it can be divided into. Besides values of energy differences, it is quite common to use enthalpy differences for processes at constant pressure in materials (section 3.3.3.7). A consequence of (2) is **Hess's law** (Atkins 1990) "The overall reaction enthalpy is the sum of the reaction enthalpies of the individual reactions into which a reaction may be divided". A **reaction enthalpy / energy** is sum of the enthalpies / energies of the bonds build minus those separated, and can be calculated from tabulated values of the bond enthalpy / energy of the different bonds without quantum mechanical calculations and fundamental energy contributions. The **bond (dissociation) enthalpy** is "the enthalpy change required to break a particular bond when the substance is in the gas phase" (Brown et al. 2015). In the same way, the **enthalpy of formation** is "the enthalpy change that accompanies the formation of a substance from the most stable forms of its component elements" (Brown et al. 2015). So, from tabulated values of the energy of the most stable forms of the elements, the enthalpy of formation (and dissociation) can be calculated in a simple way. The example in Fig. 5-4 has been discussed in section 3.2.6.2.

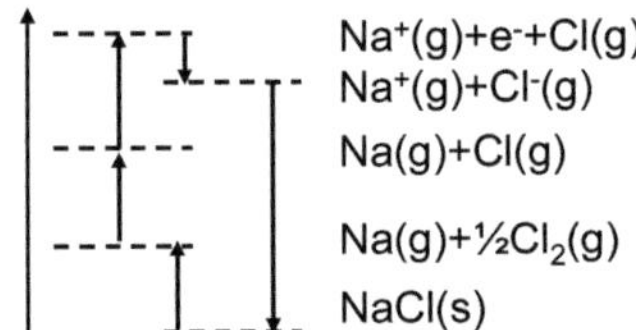

Fig. 5-4 Energetic relationship of the different process steps in the formation of NaCl (based on Brown et al. 2015)

The most basic of all is the **binding energy**, which is the energy needed to separate all the particles of a system, and thus just an energy difference between a bound state and the separated state which is at r = ∞ and also v = 0. It applies even to a nucleus composed of its nucleons.

5.3.2 Energy exchange between systems

What about the exchange of energy between systems? We can describe the interaction of two systems, affecting their external or internal energy in any of the fundamental energy contributions, by the fundamental interactions. For example for internal energy, a net electric charge of a system (from its individual charges) causes a force on the electric charges in another system, affecting their position and motion and thus the external as well as internal energy of that system. And the collision of two moving charges, affecting each other by the electric interaction, can also be described by the corresponding force. In chapter 3.1 we discussed examples for all fundamental interactions. Discussing energy exchange between systems by the fundamental energy contributions is again the easiest way to understand things.

The fundamental interactions also cover everything; no part can be missing, and no parts might be counted more than once. However, the exchange of energy is described by different concepts depending on a given situation. The most common is by a corresponding force or a force field. If looking at the time needed to exchange energy, especially to transport it across a distance, we often use waves describing the time development of a force field. And to describe energy exchange by quanta, needed if quantum mechanical effects are relevant, the force field is quantized by introducing exchange particles like the photon or graviton.

Disadvantages show up again if a situation is complex, if we don't have detailed information, if we aren't interested in details, and if QM effects exist. As we cannot see microscopic effects, and are often not interested in them, this of course applies also to the way of energy exchange between systems. We cannot explain or describe the collision of two balls, or how a ball is supported by a shelf, by the fundamental interactions directly. And we can also not describe thermal contact and associated energy exchange this way. And if two masses interact via a spring, we are also not interested in how. We don't see the details, and even scientists aren't interested in these cases. Therefore, the fundamental energy contributions are widely applied to describe the energy exchange between systems in macroscopic observations regarding the objects, their motion and position, and the way they interact and thereby exchange energy, especially by gravitation and, in limited cases, electric and magnetic interaction. Scientist use them at all scales of magnitude, but even scientists do not use them in all cases.

5.3.3 Energy conservation and conversion

Discussing energy using the fundamental energy contributions relates directly to the fundamental interactions (or corresponding forces) between the individual objects that interact, only by their relative motion and position.

Energy conservation in a single interaction is, using fundamental energy contributions, a result of the symmetry of the effect of interactions between objects on the objects. In the concept of forces this can be derived from Newton's 3rd law; forces act on both interacting objects the same distance. And since the fundamental energy contributions are "fundamental", no energy contribution can be missing, and none can be counted more than once. Thus, it is also easy to write down a correct and complete energy balance.

Energy conversion is, using fundamental energy contributions, also easy to understand. Kinetic energy is related to the motion of an object, and potential energy, meaning the interaction energies, by the relative position. Any interaction, resulting in a change of the motion of an object or the relative position between two objects, changes the relative position to other objects, thus the related interaction energy. What changes in energy conversion is how we take notice of it from the motion and position of interacting objects. Actually, saying energy is "converted" between fundamental energy contributions is misleading, because energy is not some "thing" that is converted. Better is "a change in the amount of energy in one fundamental energy contribution leads to an opposite change in another fundamental energy contribution", referring to the origin in single interactions.

Disadvantages in energy calculations show up if a situation is too complex, for example if we do not have detailed information to do the calculations, if we aren't interested in details so we do not want to do detailed calculations, and if QM effects exist. And this does not only apply to energy but also to entropy which gives information about the direction of energy conversion.

The entropy can be calculated by $S = k \cdot \ln\Omega$, with Ω being the number of accessible states of a system (Eq. 3-46). Reif 1985 says "Indeed, if one knows the nature of the particles constituting the system and the interactions between them, then one can in principle use the laws of mechanics to compute the possible quantum states of the system and thus Ω." Therefore, knowing all the detailed information we could also calculate the number of accessible states of a system, and then the entropy, at least in principle.

5.3.4 Summary and conclusions

Using fundamental energy contributions, the choice of the system is free, and related to it what is external and what is internal energy. Thus, external and internal energy both have the same fundamental energy contributions: kinetic energy, and potential energy due to the fundamental interactions, which are gravitation, the electric and its relativistic counterpart the magnetic interaction, and the strong and the weak nuclear interaction (Fig. 5-5). A system can have energy of any of the fundamental energy contributions, and it can exchange energy of any of the fundamental energy contributions by the fundamental interactions. This shows Fig. 5-5 in an overview.

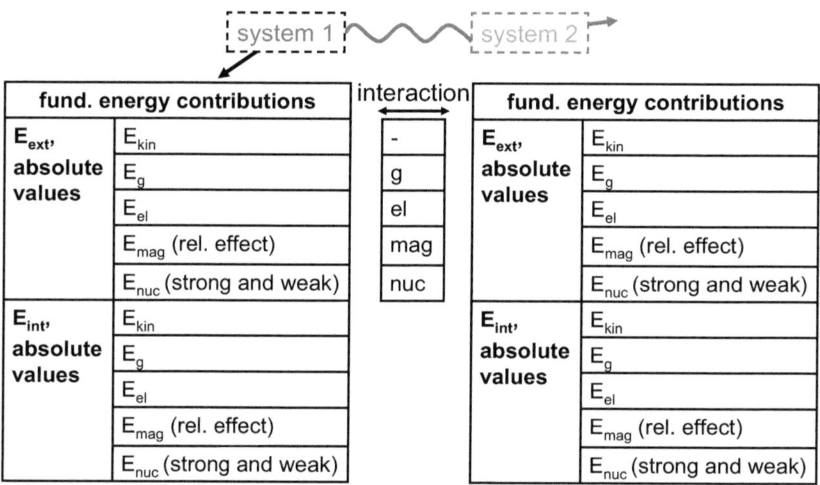

Fig. 5-5 Concept of energy using fundamental energy contributions – energy of two systems, and energy exchange between them

Discussing energy using the fundamental energy contributions relates directly to the fundamental interactions and corresponding forces. It allows a detailed treatment of the individual objects that move and that interact, only by their motion and relative position, and changes thereof; nothing else. Discussing energy using the fundamental energy contributions is thus the easiest way to understand what energy is, how it is exchanged, why it is conserved, and how it is converted. And the fundamental energy contributions cover everything; no part can be missing or counted more than once. Strictly speaking this holds within the stage chosen for the discussion here.

Disadvantages show up if a situation is complex, if we don't have detailed information, if we aren't interested in details, and also if QM effects exist. Detailed information means information on the objects that move or the objects that interact, their relative motion and position, and changes thereof.

Regarding energy of systems, the fundamental energy contributions have absolute values, usually normalizing kinetic energy to zero at zero velocity, and the potential energies at infinite distance where the interactions vanish. We can see macroscopic objects in motion, or interacting by gravitation; then, kinetic energy and gravitational energy are not only fundamental energy contributions but also energy forms. For microscopic objects, especially processes in materials on the atomic and molecular scale, the description by the fundamental energy contributions is commonly used in science.

Regarding energy exchange, there are many cases where we can observe energy exchange between macroscopic systems by gravitation, and by electric or by magnetic interaction. Thus we have corresponding types of work, meaning a force acting for a distance. On the microscopic scale, the fundamental energy contributions are again commonly used in science, however often the respective exchange particles are used to deal with quantization.

Regarding energy conversion and conservation, the fundamental energy contributions, by looking at the individual objects and their relative motion and position, are the best way to understand what energy is, how it is converted, and why it is conserved. They show that energy is not a "thing", but a concept to describe the effect of the interaction of objects on the objects. And they show it independent of the size or number of objects.

Using the fundamental energy contributions is clearly the easiest way to understand the concept of energy, its conservation, conversion, and exchange. And especially in simple situations, it is also the easiest way to apply it. However, for complex situations with many objects and several fundamental interactions, like in materials, it is not feasible using fundamental energy contributions. We do not have the detailed information, we often do not want to deal with it, and detailed calculations might also be too complex. Because the fundamental energy contributions often do not relate to things we observe directly, they are also not the way how the whole concept of energy developed in history; it was instead by energy forms. Many of them were initially not understood regarding their origin. This way of the historic development has still consequences, so let's have a quick look at it now.

5.4 Concept using energy forms

Historic development

In the beginning of natural sciences, the description of the interaction and motion of things was by forces. Later, momentum and impulse were added. The term energy, already introduced by Aristotle, came in use much later when it was realized that it allows to describe things that can't be described by forces, or momentum and impulse. First investigations related to energy, more precise the formulation of quantitative relations on conversion and conservation, started with observations on simple systems of everyday use and of macroscopic size like a pendulum that swings over and over again.

Energy forms are contributions to energy that relate to observations at a macroscopic scale. The first ones were kinetic energy of translation and rotation, and gravitational energy of macroscopic objects, all external energy. These describe the swing of a pendulum, the flow of fluids, in general the macroscopic visible motion of macroscopic bodies exposed to gravitation. Another one is related to the macroscopic visible deformation of objects. Deformation comprises changes of size and shape (like length and volume change, shear, torsion, and bending) of macroscopic objects like a spring exposed to an external force or change of force. In contrast to the motion of macroscopic bodies or their interaction by gravitation, deformation is a form of internal energy; its real origin inside the deformed body is invisible.

Macroscopic bodies are composed of small particles, invisible to the eye, such that it is not possible to observe their motion and position directly. This is why, regarding what is visible, energy seems to disappear e.g. by friction, or come out of nothing e.g. by combustion. Thus, everyday observations seem to imply that energy can be created or destroyed.

Careful measurements showed however that energy is instead conserved. The invisibility of the microscopic effects has then led to the conclusion that energy is some "thing" that cannot be destroyed, but can change appearance, thus the different energy forms. Related is the idea that heat "flows". These ideas, originating from the observation of macroscopic effects, were introduced by early scientists at a time when the microscopic basis, meaning microscopic particles, was still not commonly accepted. And the ideas are still common today among non-scientists. In fact, the misleading wording that was initially introduced is still widely used even among scientists.

With the investigation of more effects in materials even more things were not understood, for example looking at chemical reactions and thermal energy, or electric and magnetic effects in materials. When people observed that mechanical energy from deformation and thermal energy can be converted, the need for mechanical energy and the availability of thermal energy increased the range of applications of energy, starting with steam engines. Investigations increased, and the field was called thermodynamics.

Thermodynamics started with what was observed macroscopically, and later microscopic models explaining the macroscopic observations were found. As a consequence, the scientific field of thermodynamics is focused on the internal energy of macroscopic bodies, or materials they are composed of, and it is subdivided into macroscopic and microscopic thermodynamics. Macroscopic thermodynamics looks at what is observed macroscopically. Microscopic thermodynamics treats materials as systems of many particles and looks at the behavior of the particles, and how it relates to what is observed macroscopically, using statistics to describe entropy and its increase. Due to its statistical basis the common term is statistical thermodynamics. For example, energy conservation was derived from many observations on macroscopic systems, by different experiments. Because there was initially no knowledge of the microscopic basis there was no way to understand it. Today, knowing the microscopic basis, its origin can easily be understood. And the same holds for thermal energy, friction, and many other effects. Further on, while the macroscopically observed changes relate only to changes of energy, a microscopic description using the fundamental energy contributions actually allows to calculate absolute values of energy.

Historically, the concept of energy grew stepwise; after new observations new energy forms were added and later explained by microscopic models. However this way the view is piecewise, and understanding is fragmented. We discussed things here from the simple to the complex systematically. Therefore, we should now be able to see and write down the whole concept.

To understand the different energy forms, and how things relate to each other, a systematic connection of macroscopic effects with the later developed microscopic models is needed. In chapters 2 and 3 we have done this. But because our goal is a systematic, clear, and comprehensive treatment, we did it not just for internal energy, and we did it independent of the size. For the concept of energy using energy forms, we need to specify things.

Choice of the system

The application of the concept of energy affects the choice of the system, and in turn the subdivision of what is internal and what is external energy. And if a system is macroscopic or microscopic has additional consequences. In contrast to energy contributions, which are anything that contributes to the overall energy, energy forms are only such contributions that are associated with things being directly observed. Since only macroscopic things can be observed directly, for the concept of energy using energy forms we generally subdivide everything (Fig. 5-3) into macroscopic and microscopic.

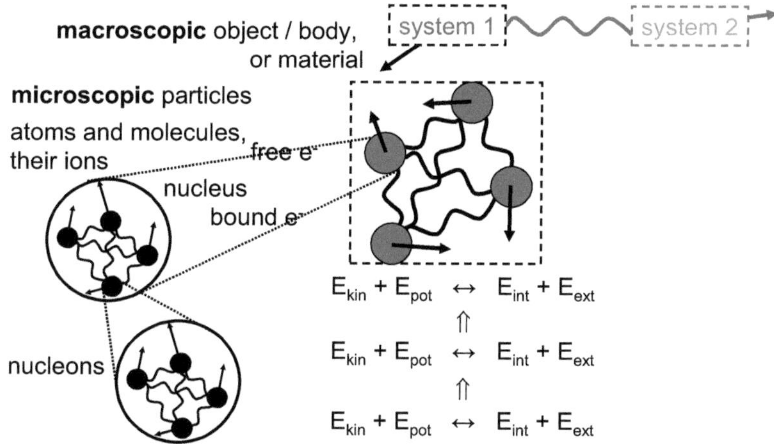

Fig. 5-6 Macroscopic system, with microscopic subsystems

The usage of the term "system" then refers to a macroscopic object / body or a material it is made of (Fig. 5-6). A system itself is composed of microscopic objects, called particles; energy referring to them is internal energy. In thermodynamics the energy of the system as a whole, its external energy, is often not discussed at all, thus assuming the system as a whole has no motion or interaction with another system, as is clear from the definitions of internal energy in literature (section 5.2.1). Most use the term "particle", and sometimes intrinsic contributions, referring to materials being composed of atoms and molecules, a **subsystem** with own energy contributions. But it is not just atoms and molecules (sections 3.2 and 3.3); it is atoms and molecules, their ions and free electrons, and if considering effects in nuclei also the nucleons and maybe even quarks, as Fig. 5-3 and Fig. 5-6 show.

The internal and external energy of a system is the sum of the internal and external energy of all its particles, including those in subsystems (Fig. 5-6). The same holds for kinetic and potential energy; it is just another way to classify all energy contributions. The kinetic energy due to a net motion is external energy, while that of relative motion (to the CM) is internal energy. The potential energy due to mutual interaction is internal energy, while that due to interaction with the surrounding, another system, is external energy. In thermodynamics the energy of the system as a whole, its external energy, is often not discussed at all, assuming the system as a whole has no motion and no interaction with another system, and thus has only internal energy. Then, the internal energy of a system is the sum of the internal and external energies of all its particles in its subsystems, and the same for kinetic and potential energies. For example, in a gas with diatomic molecules an energy contribution to the thermal energy, an energy form of the gas, is the vibrational energy of the molecules. The vibrational energy in turn comprises kinetic and potential energy of the atoms in the molecules, with the potential energy being due to electromagnetic interaction in the interatomic bonds.

Now, let us look more detailed on the consequences of the system being macroscopic, specifically how we describe things. To describe what is going on in detail we need detailed information, which means information on

- the objects that move or the objects that interact,

- the motion and relative position, and changes thereof

For microscopic objects we do not have this information from direct observation. Because macroscopic objects are composed of microscopic objects, called particles, we can observe however things related to their motion and relative position or changes thereof, even if we can not see the particles themselves directly, as changes of the macroscopic system e.g. in its volume, shape etc.

For macroscopic objects we have this information by direct observation. Energy that refers to macroscopic objects or bodies (our system, Fig. 5-6) as a whole is thus what we call external energy. It comprises energy related to the motion of the system, the macroscopic object or body as a whole, and energy related to its position by an interaction with another system.

Let's now get specific and look at the different individual energy forms, thus at macroscopic systems, first for external and then for internal energy.

5.4.1 Energy of systems – internal and external

5.4.1.1 External energy

Energy related to a single fundamental energy contribution

Kinetic energy

As external energy, kinetic energy is associated with the motion of a whole system, specifically its center of mass, and described by its velocity. The system can be anything macroscopic, like a stone, a car, or the wind.

Gravitational energy

As external energy, gravitational energy is associated with the interaction by gravitation between the mass of a whole system and that of another system, affected by their distance. As gravitation is fairly weak, at least one of the systems interacting is usually huge, e.g. a stone interacts with the earth.

Electric energy

As external energy, electric energy is associated with the electric interaction between the electric charge of a whole system and that of another system, affected by their distance, and for dipole moments also their orientation. It is rare in macroscopic systems, except e.g. fluff sticking to clothes.

Magnetic energy

As external energy, magnetic energy is associated with the magnetic interaction between the mag. dipole moment of a whole system and that of another system, affected by their distance and orientation. It is relevant on a macroscopic scale if a significant magnetic field is created or by a whole planet, e.g. a compass needle in the magnetic field of the earth.

~~"Nuclear energy"~~

Due to its short range, there is no effect on macroscopic systems.

Energy related to combined fundamental energy contributions

A system can also have external energy associated with combined fundamental interactions with another system, e.g. if a stone interacts via a spring with a wall. Deformation of the spring is based on interatomic bonds, thus associated with a combination of fundamental energy contributions.

5.4.1.2 Internal energy

External energy of macroscopic systems is, as we just saw, still largely described by the fundamental energy contributions, which relate directly to the fundamental interactions and their corresponding forces. This is possible as we can see the individual, macroscopic systems, their position and motion. Discussing internal energy the same way is however not possible as we do not have information on the position and motion of microscopic particles in a system. Thus, we need a way to relate what we observe on a macroscopic scale to internal energy. Obviously, if nothing changes macroscopically we get no information at all. And from macroscopic changes we can only derive changes of internal energy, not its absolute value. This is also obvious. If there are changes, observable macroscopically, we could start with

$$
\begin{aligned}
dE_{int} &= \left(T \cdot dS - p \cdot dV\right) + \left(E \cdot dP + B \cdot dM\right) + \left(\mu \cdot dN + ...\right) \\
&= \left(dE_{th} + dE_{def}\right) + \left(dE_{el} + dE_{mag}\right) + \left(dE_{chem} + ...\right) .
\end{aligned}
$$

Eq. 5-10

The top line is based on the description of energy exchange, but we will see in section 5.4.2.1 why we can use it here this way. It comprises three groups of terms, marked by (...), and shows many things that we have found during the discussion in chapter 3. It relates energy to things not too complex, observable, and interesting, e.g. dV or dN, and even affected by quantum mechanics, e.g. μ, in a way we can't see. What we discussed in section 3.1, where we discussed kinetic energy, gravitational energy, electric energy and magnetic energy, energy of electric, magnetic and electromagnetic fields, and finally "nuclear energy", is partly in the 2nd group and partly not listed; several things seem to be missing. What we discussed in section 3.2, where we discussed excitation energy and ionization energy, chemical energy, and nuclear energy, is in the 3rd group, also only partly. And what we discussed in section 3.3, where we discussed deformation energy and thermal energy, is in the 1st group, completely.

This raises a lot of questions. To answer them, we first compare the different groups in Eq. 5-10 with what we discussed in sections 3.1, 3.2, and 3.3. Afterwards, we will try to discover the systematics that leads to Eq. 5-10. Specifically, we will try to find what is characteristic for the individual terms, and check if nothing is missing or counted more than once. And finally, once we understood the systematics and checked things, we will of course try to give clear and precise definitions of the different energy forms.

Let's now start comparing the different groups in Eq. 5-10 with what we discussed in sections 3.1, 3.2, and 3.3. We begin with the 1st and 2nd group, which relate to changes without a change of the composition of particles.

Changes without a change of the composition of the particles

The 2nd group, ($dE_{el} + dE_{mag}$), relates to what we discussed in section 3.1. The internal **electric energy** is the energy due to the electric interaction between the charges in a system, affected by their relative positions. In general, we do not have information on the individual positions, nor do we want to know them or deal with them. But for a special case, the directed separation of opposite charges, there is a simple way to describe the energy macroscopically from microscopic information, like in a capacitor. The reason for a change can be an electric field from the surrounding that causes a directed force on all charges, resulting in a **directed change of the relative positions**. Macroscopically this is observed by a change in the overall polarization P, and related to a change of the internal electric energy by E·dP. The term B·dM describes the equivalent for a magnetic field B and the change of the magnetization M, resulting in a change of **magnetic energy**. It is necessary to stress that E·dP refers to all electric charges and dipoles, respectively B·dM to all magnetic dipoles, in all subsystems of the system. We also discussed in section 3.1 why there is no term for gravitation if the field is homogeneous, and besides, on a microscopic scale the contribution of gravitation to energy is negligible anyway. There is also no nuclear part; the nuclear forces have such a short range that they only act within nuclei. Why is there no part for kinetic energy? A **directed change of the motion** of all particles means a net motion of the whole system, thus external energy. And a random, **undirected change**, shows up in what comes next.

The 1st group, ($dE_{th} + dE_{def}$), relates to what we discussed in section 3.3, the energy related to the relative motion and position of particles in a system: whole atoms, molecules, their ions and free electrons, however not of their subsystems meaning electrons and nuclei in atoms, in molecules, or in ions; such cases are in the 3rd group. The term p·dV, or F·ds, macroscopically describes deformation, microscopically a **directed change of the relative motion and position**, related to **deformation energy**. The term T·dS relates to a random, **undirected change of the relative motion and position**, and thus **thermal energy**. It is important to stress that "undirected" refers to the changes, which then finally result also in undirected motion of the particles.

Changes of the composition of the particles

We have now discussed all options of directed or undirected changes of the position and motion of particles. Left are changes of the particles themselves. They can all be combined in the 3rd group, $(dE_{chem} + ...)$ in Eq. 5-10.

In section 3.2 we discussed atoms, consisting of a nucleus and a number of electrons orbiting around it, and saw that the situation is stabilized as the energies are quantized due to quantum mechanical effects. The same holds for molecules, an in a similar manner for the nucleus consisting of nucleons. A change of composition usually refers to the formation of "new" particles. For atoms and molecules, this is by ionization leading to ions and free electrons or reverse, and by chemical and nuclear reactions leading to new chemical compounds and new nuclei. The related changes of energy are the **ionization energy**, the **chemical reaction energy**, and the **nuclear reaction energy**. Ionization of atoms or molecules is an effect that is usually not mentioned, and for nuclei there is no corresponding effect at all. Observable on a macroscopic level are e.g. chemical reactions if the material properties change in an observable way, and ionization e.g. by a change of the electrical conductivity of the material. For nuclear reactions, the directly observable effect is e.g. the energy difference. In all cases, the energy is not an absolute energy but an energy difference ΔE between two different situations. To simplify its calculation e.g. the energy of formation, binding energy, or bond energy can be used (Fig. 5-4). The description for the bulk is by $dE_{...} = \Delta E_{...} \cdot dN$, where ΔE refers to the energy change of a single particle. This approach can also be used if the composition changes in a solution, being the other type of chemical process besides chemical reactions.

Changes of composition can also be without the formation of new particles, just by a minor rearrangement. For atoms there is **excitation energy**, and the same for molecules and ions. In a similar way, nuclei, consisting of one or several nucleons, can also be excited and thus have **excitation energy**. Observable on a macroscopic level is the excitation of atoms and molecules commonly by absorption or emission of light (photons, Fig. 3-35). Again, the energy has not an absolute value but is an energy difference ΔE between two different situations, thereby avoiding to bother with what goes on at the microscopic scale. The description for the bulk is by $dE_{ex} = \Delta E_{ex,1} \cdot dN$, similar to above, where $\Delta E_{ex,1}$ refers to the energy change of a single particle.

Is anything missing? Is anything counted more than once?

The just discussed set of energy forms clearly has a **systematic structure**. But is it complementary, nothing missing, nothing counted more than once?

The energy forms look only at systems that are macroscopic, but that does not mean that they miss something. Microscopic changes related to energy are included by related macroscopic observations, e.g. a change of volume. The discussed set of energy forms is within the set stage, covering objects in motion and interacting with each other. It does also comprise energy of electromagnetic fields implicitly by the charges that cause them, and even energy associated with a change of mass by $E = m \cdot c^2$ in nuclear reactions.

A macroscopic system has **external energy** as kinetic and potential energy, by its motion as a whole and its interaction as a whole with another system. There is, as energy related to just a single fundamental energy contribution **kinetic energy, gravitational energy, electric energy, magnetic energy**, but not nuclear energy; a macroscopic system does not interact with another system by the nuclear interactions due to their limited range. Obviously, this is already a complementary set if just looking at what contributes to energy. It is simply by the fundamental energy contributions, with absolute values. However, regarding a convenient description by what is macroscopically observed, energy related to **combined fundamental energy contributions** should be added, like the interaction by a spring with a macroscopic change of length. The related energy is deformation energy of the spring, but from the viewpoint of an attached mass as the system, the mass has external energy due to the interaction via the spring with something else, e.g. a wall.

A macroscopic system has **internal energy** as kinetic energy by motion of its microscopic particles relative to the CM, and potential energy by the mutual interaction of all particles, affected by their relative position. Motion that is not relative to the CM is already accounted for as external kinetic energy, and interaction that is not mutual, thus with the surrounding, is already accounted for in external potential energy. The energy forms do however not refer to the relative motion and position, but to things observable on the system on a macroscopic level, like dV. To see if the energy forms of internal energy are complementary we thus have to check the microscopic basis. We already split internal energy by energy associated with "Changes without a change of the composition of the particles" and energy associated with "Changes of the composition of the particles"; let's start with the last one.

249

The 3rd group in Eq. 5-10, $(dE_{chem} + ...)$, comprises changes of internal energy that are associated with "Changes of the composition of the particles". A **particle** is here an entity that can be composed of even smaller parts; however, they must stay together as an entity for a "significantly" long time. The reason is that in microscopic particles quantum mechanical effects cause discrete energy levels, such that only if the energy difference can be exchanged with the surrounding a change within the particle is possible. Such particles are nuclei of atoms, atoms being composed of a nucleus and electrons, and molecules being composed of atoms. Changes of the composition can be essential and thus lead to a new particle, or they just lead to an energetically excited state. Such **excitation energy** exists for atoms and molecules if electrons change their orbit, and for nuclei by their nucleons. Essential changes of composition that lead to a new particle are for atoms and molecules ionization, leading to ions and free electrons or reverse, and chemical reactions. Related is **ionization energy** and **chemical reaction energy**. For nuclei it is nuclear reactions and **nuclear reaction energy**. There is nothing corresponding to ionization for nuclei; a loss or gain of a nucleon or electron is considered a reaction as a new nucleus results. This discussion shows that all options are covered, and nothing is double; for "Changes of the composition of the particles" the set is complementary.

Now, what about energy associated with "Changes without a change of the composition of the particles". We have split this already into two groups.

The 2nd group, $(dE_{el} + dE_{mag})$, relates to what we discussed in section 3.1: energy of all particles by one of the fundamental energy contributions. We discussed why kinetic energy from the relative motion is no energy form, and also why the gravitational and nuclear interactions do not contribute. The term E·dP describes the change of the internal **electric energy**, which is the energy due to the electric interaction between el. charges affected by their relative distance or el. dipoles in addition affected by their rel. orientation. The term B·dM describes the equivalent for a magnetic field B, magnetization M, and **magnetic energy**. This discussion again shows that all relevant options are covered, and that nothing is counted more than once.

We still have another group not discussed yet, the 1st group. Why actually? We covered all options for changes within all particles, and we covered all options for single fundamental energy contributions of all particles. Is there something that is still missing, and that is covered by the 1st group?

The 1st group, $(dE_{th} + dE_{def})$, relates to what we discussed in section 3.3. The term $p \cdot dV$ macroscopically describes deformation, microscopically a **directed change of the relative motion and relative position**, related to **deformation energy**. We already discussed that kinetic energy of random motion is included in the 1st group somehow. What is striking is however that there is again energy associated with changes of the relative position, actually directed changes, like in the 2nd group. But there is a difference. The 2nd group contains electric and magnetic energy by a directed change of the relative locations of charges and orientation of dipoles such that a net polarization respectively magnetization results. When an ionic crystal is deformed, all charges (positive and negative) change position such that no net polarization results. This is why the 1st group comprises something new. The term $T \cdot dS$ relates to random, **undirected change of the relative motion and relative position**, and thus **thermal energy**. It refers to the same particles, atoms and molecules, their ions, and also free electrons, however the changes are made in a random way. It is clear that the directed and the undirected changes in the terms $p \cdot dV$ and $T \cdot dS$ are complementary.

We have now discussed the energy forms of internal energy and of external energy, and can finally check if they all are together a complementary set. Knowing the difference between the terms $p \cdot dV$ and $E \cdot dP$, even if the particles are the same el. charges like in an ionic crystal, we know that the terms in the 1st and 2nd groups comprising the energy forms of internal energy by "Changes without a change of the composition of the particles" are a complementary set. They are also complementary to those related to "Changes of the composition of the particles", thus all the energy forms of internal energy are a complementary set. And these are complementary to the energy forms of external energy. Thus, the energy forms discussed here are a complementary set; nothing is missing, and nothing is counted more than once. Trying to find out if the set of energy forms is a complementary set shows that some terms like energy of ionization and of excitation are often not accounted for. And it helps to discover the specifics of each energy form.

We can test our knowledge by discussing some effects not discussed yet. The energy of a phase change between solid, liquid, and gas phase relates to random, undirected change of the relative motion and position of the particles, accompanied by a change in entropy; it is thus just a special case of a change of thermal energy. If the volume also changes against an external pressure, e.g. ambient pressure, then deformation energy also plays a role.

And when a substance A dissolves in a substance B there is an effect on the composition via the concentration, and also on entropy, thus dN and dS. Energy of solution therefore comprises thermal as well as chemical energy. Phase change energy and energy of solution could also be called forms of energy, but they are based on the complementary set that we discussed here.

5.4.1.3 Definitions of energy forms

We have now compared the different groups in Eq. 5-10 with what we discussed in sections 3.1, 3.2 and 3.3, we have also discovered the systematics in Eq. 5-10, specifically the characteristics of the different terms, and we found that no energy is missing or counted more than once. Now, it is time that we try to find clear and precise definitions for the energy forms.

The energy forms are quite practical, but the set of energy forms, developed throughout history just from macroscopic observations and experience, is hard to understand in detail. Scientists can avoid this by using fundamental energy contributions in their calculations. However, for common applications we need energy forms. What can we do to understand their individual meaning better? The energy forms of internal energy cover energy changes, not absolute energies, and they are relating energy changes to effects that are observed macroscopically, e.g. a change of volume, of polarization, of amount of a substance etc., instead the position and motion of the microscopic particles. This is due to a lack of detailed information on position and motion as it is not seen, because it is not interesting or too complex.

In order to get the general idea, we look at changes of internal energy from all energy forms by Eq. 5-10 again, but reorganizing the terms optically

$$
dE_{int} = \begin{pmatrix} + T \cdot dS \\ - p \cdot dV \end{pmatrix} \atop \begin{pmatrix} + E \cdot dP \\ + B \cdot dM \end{pmatrix} \atop \begin{pmatrix} + \Delta E_{chem,1} \cdot dN_{chem} \\ + \Delta E_{nuc,1} \cdot dN_{nuc} \\ + \Delta E_{ex,1} \cdot dN_{ex} \\ + \Delta E_{ion,1} \cdot dN_{ion} \end{pmatrix} = \begin{pmatrix} + dE_{th} \\ + dE_{def} \end{pmatrix} \atop \begin{pmatrix} + dE_{el} \\ + dE_{mag} \end{pmatrix} \atop \begin{pmatrix} + dE_{chem} \\ + dE_{nuc} \\ + dE_{ex} \\ + dE_{ion} \end{pmatrix}
$$

Eq. 5-11

Because the energy forms are hard to understand in detail, when trying to define them we will follow closely their mathematical formulation to avoid errors and ambiguities.

A crucial point is that when describing internal energy using energy forms we deal with energy changes, not absolute values as when using fundamental energy contributions. Missing detailed information is a general reason for this; if we do not see details, we have only macroscopic changes that relate to energy, and thus changes of energy. In addition, when a composition changes we also only describe changes between different states. But there is something else that is crucial to understand energy forms of internal energy.

We can rewrite Eq. 5-11 in a general mathematical description, i denoting the different energy forms, using basic differential calculus (e.g. Eq. 3-58)

$$dE_{int} = \sum_i \left. \frac{\partial E_{int}}{\partial X_i} \right|_{X_{j \neq i} = const} \cdot dX_i = \sum_i Y_i \cdot dX_i \qquad \text{Eq. 5-12}$$

Because of the resemblance to $dE = F \cdot ds$, the Y_i are commonly called **generalized force** and the dX_i the corresponding **generalized displacement**. All terms of energy forms of internal energy are written as what is called an extensive and an intensive variable. An **extensive variable** scales with the size of a system, while an **intensive variable** does not scale with it. The extensive variables, being S, V, P, M, N_{chem}, N_{nuc}, N_{ex}, and N_{ion}, are the crucial variables that helped us to discover that this is a complementary set. They are the generalized displacements. The intensive variables T, -p, E, B, $\Delta E_{chem,1}$ ($=\mu$), $\Delta E_{nuc,1}$, $\Delta E_{ex,1}$, $\Delta E_{ion,1}$ are the generalized forces, which we need to calculate the related change of energy. We already discussed in section 3.3.3.5 that the partial derivatives of E_{int} with respect to the extensive variables are related via the Maxwell relations. They can therefore change in a process due to a change in another energy form. This is why in Eq. 5-12 the partial derivative of E_{int} with respect to X_i is with all other $X_{j \neq i}$ constant. The most crucial examples are probably thermal and deformation energy. The value of p can change if T changes, thus also $p \cdot dV$. And the value of T can change if p changes, thus also $T \cdot dS$. This means that for changes of S or V which are not differential, $T \cdot dS$ and $p \cdot dV$ depend on the conditions along the path (section 3.3.3.5). This is why thermal engines can do work. For specified conditions we can integrate the changes, e.g. for V = const. integrate $T \cdot dS$ and get a value of ΔE_{th} that is valid just for these conditions.

Characteristic of an energy form is thus the change of its extensive variable, and the related intensive variable at conditions specific for a given situation. With this knowledge in our mind, we will now try to make clear definitions, and we will also compare them with definitions found in literature.

Changes of the composition of the particles

Nuclear reaction energy

Nuclear reaction energy is the energy (internal energy) that is associated with a change of the number of nuclei of a specific type, and due to the energy difference of the nuclei before and after a change of composition in a nuclear reaction. Its basis are the strong and weak nuclear interaction between the nucleons (precisely the quarks), also affected by their electromagnetic interaction. Examples of nuclear reactions are fusion, fission, or decay; excitation might be included.

Cleveland and Morris 2009 state that nuclear energy is "energy released by radioactive decay, through a nuclear reaction, or in the course of fission (splitting) or fusion (fusing) of atomic nuclei". Here, the physical basis is missing, and actually decay, fission, and fusion are all examples of nuclear reactions. Çengel and Boles 2002 state "The tremendous amount of energy associated with the strong bonds within the nucleus of the atom itself is called nuclear energy." Similarly, Çengel 1998 states that "The internal energy associated with the bonds within the nucleus of the atom itself is called nuclear energy." Both refer to the bond energy, not the reaction energy.

Chemical reaction energy

Chemical reaction energy is the energy (internal energy) that is associated with a change of the composition of chemical compounds, thus of the arrangement of atoms in molecules, and due to the energy difference of the compounds before and after a change of composition in a chemical reaction. Its basis is the energy of bonds broken or built, which is due to the electromagnetic interaction between atoms and electrons and their kinetic energy. Examples of chemical reactions are decomposition and oxidation.

Cleveland and Morris 2009 state that chemical energy is (chemistry) "energy produced or absorbed in the process of a chemical reaction". This is correct, but not very specific. Çengel 1998 and Çengel and Boles 2002 state "The internal energy associated with the atomic bonds in a molecule is called chemical energy." But this is actually the bond energy, and just refers to atomic bonds. And it does not state what the energy specifically refers to.

Ionization energy

Ionization energy is the energy (internal energy) that is associated with a change of the number of electrons (a gain or a loss) in the electron shell of atoms or molecules, and due to the energy difference between before and after an electron is lost or gained. Its basis is the electromagnetic interaction between the atoms and electrons and their kinetic energy.

Brown et al. 2015 state that ionization energy is "the energy required to remove an electron from a gaseous atom when the atom is in its ground state". This gives specific boundary conditions, limiting the meaning; it could also be molecules, it does not have to be the ground state, and it does not have to be in the gas phase. And it is also not only for removal. Atkins 1990 states "The ionization energy is the change in internal energy for the same process at $T = 0$." The definition of the process was before as "removal of an electron". Again, $T = 0$ is an unnecessary limiting boundary condition.

Excitation energy

Excitation energy is the energy (internal energy) that is associated with an electron changing its energetic state in the electron shell of atoms or molecules, or with the nucleons in a nucleus, and due to the energy difference between before and after the change. Its basis is the electromagnetic interaction between the atoms and electrons and their kinetic energy in atoms or molecules, or the nuclear interactions between the nucleons in nuclei.

No definition in literature was found.

Changes without a change of the composition of the particles

Thermal energy

Thermal energy is the energy (internal energy) that is associated with a change of the entropy, meaning the randomness, in a many-particle system. It is due to the energy difference between before and after the entropy change, and affected by the temperature. Its basis is the kinetic and potential energy of atoms and molecules, their ions, and free electrons, more precise changes of the rel. motion and position related to randomness / disorder.

Cleveland and Morris 2009 state that „thermal energy" is (thermodynamics) "the kinetic energy associated with the motion of atoms or molecules in a substance; i.e. heat". This is wrong in several ways. First, it is not just atoms and molecules, but also ions and free electrons. Second, it misses potential energy, e.g. the vibrational contributions of molecules. But more important, it misses the crucial point, the change associated with a change of entropy. And "heat" is wrong as it denotes energy exchanged between systems and not energy of systems. Halliday et al. 1993 write "This energy is internal energy (or thermal energy), which is the collective kinetic and potential energies associated with the random motions of the atoms, molecules, and other microscopic bodies within an object." This sounds better, but thermal energy is not the kinetic and potential energies associated with the random motions as we saw; this would have an absolute value, and thus the difference between two states would not be dependent on the path. Instead, thermal energy is the energy change associated with a change of the entropy. These differences are crucial for understanding the basics.

Deformation energy

Deformation energy is the energy (internal energy) that is associated with a change of the form of a many-particle system, meaning its size or shape, thus a directed, ordered change. It is due to the energy difference between before and after the change, e.g. for volume changes affected by pressure. The force needed to compress a spring depends on its initial state; it becomes larger the more a spring is already compressed. And when air is compressed the pressure is easily modified by its temperature, which shows the importance of the conditions. Basis of deformation energy is the kinetic and potential energy of atoms and molecules, their ions, and free electrons, more precise the change by an ordered change of their motion and position.

The following energy forms refer to internal energy and also to external energy (discussed in section 5.4.1.1.); they are thus discussed here together.

Electric energy

Electric energy is the energy of an array of charged particles or dipoles because of the electrostatic interaction, affected by their relative location (distance) and for dipoles also by their relative orientation.

As external energy, electric energy is associated with the electric interaction between the electric charge of a whole system and that of another system, affected by their distance, and for dipole moments also their orientation. It is rare in macroscopic systems, except e.g. fluff sticking to clothes.

As internal energy, electric energy is the energy that is associated with the electric interaction of el. charges and dipoles in a system, affected by their distance and orientation. Macroscopically observable is the change of internal energy by a change of polarization of the system at a given electric field.

Cleveland and Morris 2009 state that electrical energy is (electricity) "(1) the energy inherent in an array of charged particles because of their relative positions. (2) the energy inherent in a circuit because of its position in relation to a magnetic field". (1) describes the physical basis, however misses to name the interaction, and (2) is wrong as this is magnetic energy.

Magnetic energy

Magnetic energy is the energy of an array of magnetic dipoles because of the magnetic interaction, affected by their relative location and orientation.

As external energy, magnetic energy is associated with the magnetic interaction between the mag. dipole moment of a whole system and that of another system, affected by their distance and orientation. It is relevant on a macroscopic scale if a significant magnetic field is created or by a whole planet, e.g. a compass needle in the magnetic field of the earth.

As internal energy, magnetic energy is the energy that is associated with the magnetic interaction of mag. dipoles in a system, affected by their distance and orientation. Macroscopically observable is the change of internal energy by a change of magnetization of the system at a given magnetic field.

Gravitational energy

Gravitational energy is the energy of an array of objects with mass because of the gravitational interaction, affected by their relative location (distance).

As external energy, gravitational energy is associated with the interaction by gravitation between the mass of a whole system and that of another system, affected by their distance. As gravitation is fairly weak, at least one of the systems interacting is usually huge, e.g. a stone interacts with the earth.

As internal energy, gravitational energy is only used when using fundamental energy contributions, but not using energy forms as it is not relevant on microscopic scales.

Çengel and Boles 2002 state that "The energy that a system possesses as a result of its elevation in a gravitational field is called potential energy." And Cleveland and Morris 2009 define gravitational potential energy (physics) as "the energy that an object possesses because of its position in a gravitational field". Both definitions do not point to the real origin, the interaction by gravitation with another mass, thus risking the misunderstanding that the energy belongs only to the system. In addition, the first definition also implies that gravitational energy and potential energy are identical.

Kinetic energy

Kinetic energy is the energy of one or several objects due to their motion, characterized by their velocity.

As external energy, kinetic energy is associated with the motion of a whole system, specifically its center of mass, and described by its velocity. The system can be anything macroscopic, like a stone, a car, or the wind.

As internal energy, kinetic energy is only used when using fundamental energy contributions, and not as an energy form because the microscopic particles in a macroscopic system and their motion is not visible.

Çengel and Boles 2002 state "The energy that a system possesses as a result of its motion relative to some reference frame is called kinetic energy". This just describes it as external energy; for that it is correct.

5.4.1.4 Overview

Let's now summarize what we found about energy of macroscopic systems as described by energy forms, and compare it with the description using fundamental energy contributions. A brief overview shows Fig. 5-7.

For external energy, the energy forms are almost identical with the fundamental energy contributions. There is kinetic energy, and potential energy by fundamental interactions as gravitational, electric and magnetic energy (nuclear is not relevant on a macroscopic scale). All have absolute values.

Potential energy can also be due to a combination, for example if a macroscopic system interacts by a spring with its environment.

fund. energy contributions	
E_{ext}, absolute values	E_{kin}
	E_{g}
	E_{el}
	E_{mag} (rel. effect)
	E_{nuc} (strong and weak)
E_{int}, absolute values	E_{kin}
	E_{g}
	E_{el}
	E_{mag} (rel. effect)
	E_{nuc} (strong and weak)

energy forms	
E_{ext}, absolute values	E_{kin}
	E_{g}
	E_{el}
	E_{mag}
	E_{comb}
E_{int}, differences	dE_{th}
	dE_{def}
	dE_{el} — (incl. $E_{el\text{-}mag}$ fields) — dE_{mag}
	dE_{chem}
	dE_{nuc} (incl. implicitly $E = m \cdot c^2$)
	dE_{ex}
	dE_{ion}

Fig. 5-7 Comparison of external and internal energy by fundamental energy contributions (left, see Fig. 5-5) and by energy forms (right)

For internal energy, the energy forms are different from the fundamental energy contributions. Historically the terms come from what can be observed, is of interest, not too complex, in short what is applied and relevant. The energy forms of internal energy, following the structure discussed here, are thermal energy and deformation energy, electric and magnetic energy, chemical and nuclear energy. To make the system complementary within the stage of the discussion, ionization and excitation energy must be added. In general, the energy forms of internal energy refer only to energy changes, which depend on conditions by the generalized forces. This is quite different from the fundamental energy contributions that refer to a given situation and give an absolute energy value. Thus, the energy forms allow to use the concept of energy by easily observable changes, but are less precise and specific than the fundamental energy contributions. The effort needed to get a complementary structure of the energy forms, and to check the individual definitions in detail, shows that energy forms are convenient, but difficult when trying to be precise, especially when describing things by words.

5.4.2 Energy exchange between systems

After the discussion of the energy of macroscopic systems by energy forms, let's now discuss the energy exchange between such systems. In energy exchange, a system interacts with another system, which can also be its surrounding, and as a result the energies of the system and the other system change.

5.4.2.1 Classification of energy exchange by interaction

In section 5.3.2 we looked at energy exchange between systems using the fundamental energy contributions. The result summarizes Fig. 5-5. We saw that energy exchange is simply described by the fundamental interactions. What changes in the concept of energy if we are now using energy forms? The result summarizes Fig. 5-8. It shows again why energy as a concept is so incredibly useful: it describes changes in so many different things in a quantitative way, and allows predictions in a wide variety of situations.

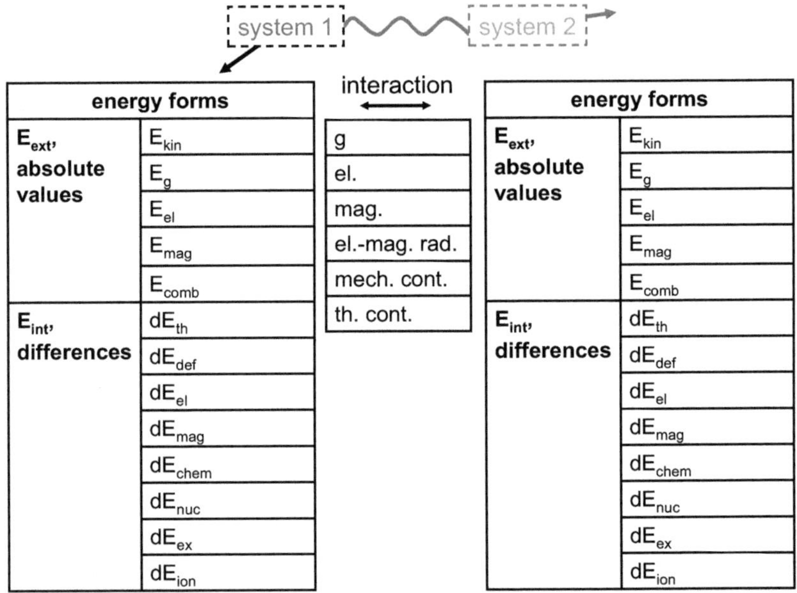

Fig. 5-8 Concept of energy using energy forms – energy of two systems, and energy exchange between them

For external energy there are again the fundamental interactions, thus gravitation, electric and magnetic interaction, but not the nuclear interaction as its range is too short to be relevant for interactions between macroscopic systems. These interactions are commonly described by the corresponding forces or their force fields. Further on, there are interactions by "mechanical contact" by hand or collisions for example, which are a way to describe the interaction of macroscopic objects without going into any details about the origin, the microscopic basis. For example, two magnets can exchange energy by magnetic interaction, e.g. rotational kinetic energy. The swing-by maneuver in Fig. 2-36 shows the exchange of kinetic energy by gravitation, while if two balls collide the energy is exchanged by mechanical contact.

For internal energy there is again the electric and magnetic interaction, e.g. the charge of a system polarizes another system changing its electric energy, or a magnet magnetizes a piece of iron thus changing its magnetic energy. Further on, electric and magnetic interaction act together in energy exchange by el.-mag. waves. When the moon and the earth interact by gravitation (section 3.1.2.4) the liquid water is set in motion, causing the tides, and even the solid earth moves; in exchange, the kinetic energy of the moon changes of course also. Because of its short range however, there is no energy exchange between macroscopic systems by the nuclear interaction, not for external and also not for internal energy. Two important types of interaction are mechanical contact, leading to the deformation of a system, and thermal contact, which is by random motion on a microscopic level. What is macroscopically observed as "mechanical" contact (by hand, collision e.g. between a ball and a wall) or thermal contact, is on a microscopic scale the short-range electromagnetic interaction between the microscopic particles like atoms and molecules at the contact surface of both systems. As discussed earlier, there is no kind of energy exchange corresponding to chemical or nuclear energy, ionization or excitation energy. These energy forms refer to changes in the particles, and energy is exchanged in these cases by one of the other ways, e.g. electromagnetic radiation or collision. Of course, energy can be exchanged together with mass, e.g. when you refuel your car, but we decided earlier to exclude this case to focus just on energy.

It is important to note that there are cases where the exchange is between internal energy of one system and external energy of the other system, e.g. a ball with kinetic energy hits a car door, resulting in deformation energy, and if two cars crash, the result is even kinetic and deformation energy together.

Let's now focus on a crucial, and often confusing issue of the concept of energy regarding energy of systems and energy exchange between systems. If a mass is lifted a height h against the force of gravitation F_g, the energy of the mass changes by $\Delta E_{ext} = F_g \cdot h$. Here $F_g \cdot h$ describes the energy change of the system, and the energy exchanged with the device lifting the mass. We have the same views regarding changes of internal energy of a system. One is looking at the energy change of the system, by Eq. 5-10 or Eq. 5-11, the second is how energy is exchanged; corresponding to Eq. 5-10 this is by

$$
\begin{aligned}
dE_{int} &= (T \cdot dS) + (-p \cdot dV + E \cdot dP + B \cdot dM) \\
&= \quad \delta Q \quad + \qquad \qquad \delta W \quad ,
\end{aligned}
$$

Eq. 5-13

and corresponding to Eq. 5-11 we can similarly rewrite Eq. 5-13 and get

$$
dE_{int} = \begin{pmatrix} + T \cdot dS \\ - p \cdot dV \\ + E \cdot dP \\ + B \cdot dM \end{pmatrix} = \begin{pmatrix} + \delta Q \\ + \delta W_{def} \\ + \delta W_{el} \\ + \delta W_{mag} \end{pmatrix}.
$$

Eq. 5-14

Eq. 5-10 describes the change of internal energy by changes in the energy forms; the intensive variables T, p, E, B, ΔE_{chem} … refer thus to the system. Eq. 5-13 describes the energy exchange with the surrounding, being by a thermal contact resulting in energy exchange of T·dS (heat), by a mechanical contact as p·dV or by an electric or magnetic field as E·dP and B·dM (all work). Here, T, p, E, and B refer to the surrounding. Books like Gaskell and Laughlin 2018, Morse 1969, and Hari Dass 2014 discuss this viewpoint, looking at energy exchange. They also have a term …·dN for energy exchange by mass, which we don't as we excluded exchange of mass in the discussion to focus on energy. Finally, why do we need the two viewpoints? Both sides can of course have the same terms, e.g. electric energy and electric work as discussed in section 3.1.3.4. However we need both if they are different, e.g. if the work is by the force of gravitation as discussed in section 3.1.7.1. And energy exchange can also affect several forms of internal energy; in section 3.1.7.2 we discussed the case of an electric work from the surrounding changing the electric and gravitational energy of two plates. And last, but not least, the energy in internal energy forms can change without any interaction with the surrounding by internal conversion, e.g. chemical energy can be converted to thermal energy in a system by a combustion.

5.4.2.2 Classification of energy exchange by work and heat

As already indicated in Eq. 3-51 and now again in Eq. 5-13 and Eq. 5-14, there is another way for classifying how energy is exchanged by a system: as work or as heat. After many observations the **1st law of thermodynamics** as **"the change of the internal energy of a system is equal to the amount of heat supplied to the system, and less the work done by the system"** was derived. We discussed it in section 3.3.3.4. It specifies how (internal) energy is exchanged by a system with its surrounding: as work or as heat. What do we need work and heat for? The just discussed ways to describe energy exchange by the specific interaction are in many cases too detailed; work and heat are less detailed, and moreover they have a specific focus. We have discussed the terms heat, and especially work, already before. Work was initially introduced by the work - kinetic energy theorem. Later it was expanded by adding potential energy to the work - energy theorem in section 2.1.3.5., as energy exchanged between an object and its surrounding, specifically by a force acting a distance. But a force times distance is not always work; a change of a potential energy is described in the same way. Later we discussed systems, and in section 3.3.3.4 combined work and heat. Internal energy can change due to work, and in addition, when referring to a many-particle system, due to heat. External energy refers to a system as a whole, like a single object. A single object cannot carry disorder and thus entropy; changes of external energy are thus only due to work, and not heat. Essentially, work and heat divide energy exchanged by an interaction into being directed / ordered, thus with no entropy, or undirected / random, thus with entropy. Together they are again complementary. We can summarize: **Heat and work denote energy exchanged between systems (or a system and its surrounding). As a consequence, heat and work are no energy forms; especially, heat is not equal to thermal energy as is often said. Work is energy exchanged without disorder and thus without entropy, by a directed force (interaction) acting a distance. Heat is energy exchanged with disorder and thus with entropy, by undirected, random forces (interactions). Energy exchange by work is driven by an imbalance of forces, while for heat it is an imbalance of temperature.**

An example where work and heat are used is the calculation of the efficiency of thermal / heat engines in section 3.3.3.6. Here, reversibility is crucial, and work and heat allow taking it into account inherently by entropy.

Discussion of the definitions of work and heat

Let's now discuss some definitions of work and heat found in literature.

Sears and Zemansky 2016 define "work" as "force times displacement". However, this is in general an energy difference and not necessarily work. And it does not state what the nature of work is. Cleveland and Morris 2009 define „work" (thermodynamics) as "the useful transfer of energy to a body by the application of a force, causing the body to move a certain distance in the same direction as the applied force (product of force and displacement)". This is however limited to a change of external energy. "Work is the energy transfer associated with the force acting through a distance", as defined by Çengel and Boles 2002, is a better definition. Atkins 1990 adds an important aspect, stating that "Work is done during a process if that process could be used to bring about a change in height of a weight somewhere in the surroundings." This indicates that there are different types of work, as shown in Eq. 5-14, and that conversion between them is possible.

Sears and Zemansky 2016 define heat by "energy transfer that takes place solely because of a temperature difference is called heat flow or heat transfer, and energy transferred this way is called heat". Similarly, Cleveland and Morris 2009 define „heat" (thermodynamics) as "a measure of the amount of energy transferred from one body to another because oft the temperature difference between those two bodies". But it is not just a measure. Çengel and Boles 2002 state that "heat is defined as the form of energy that is transferred between two systems (or a system and its surroundings) by virtue of a temperature difference." Here, the use of "energy form" is misleading.

In general, these definitions refer to energy "transfer"; we used "exchange" because we found that energy is not some "thing". The definitions also state that work is related to a force and a displacement, and that the cause of heat transfer is a temperature difference. They miss however that the cause of work is a difference of forces, and that entropy corresponds to the distance. Entropy is crucial to discuss reversibility of processes, but there is more. The focus of thermodynamics on materials has left out the treatment of electromagnetic radiation. It should be included in Eq. 5-13 and Eq. 5-14, but is neither due to a net force nor to a temperature gradient (heat conduction). Sometimes in literature it is given as a third way besides heat and work. More consistent is however to define work and heat by entropy; then radiation can be included directly depending on how much entropy it carries.

5.4.3 Energy conservation and conversion

Because of the invisibility of what goes on on a microscopic scale, the energy forms treat only observable macroscopic changes of systems. The energy forms of internal energy, describing energy of systems, are thus hard to understand in detail, compared to the fundamental energy contributions. This also holds for the description of the energy exchange between systems. And it makes it hard to understand conversion and conservation of energy. How can we approach these last, but very crucial points of our discussion? Our systematic analysis has equipped us with three things:

1) The fundamental energy contributions cover everything, and in a way that we can understand it by the motion of objects and their position related to mutual interactions. Here we understand energy conservation as well as conversion.

2) We managed to relate the energy forms to the fundamental energy contributions, so we understand the basis of each energy form.

3) We have shown that the energy forms are a complementary set, so nothing is missing and nothing is counted more than once.

Energy conservation is then again a logic consequence of these three points. Because the energy forms, as discussed here, are a complementary set, within their framework energy is conserved as before for fundamental energy contributions. As a result, the changes in all energy forms cancel out, meaning the sum of all energy changes is zero (Eq. 5-8 and Eq. 5-9). It is not possible to see this directly looking at the energy forms, and this is why energy conservation is historically just an outcome of experiments. Only by the combination of above three points it is possible to understand its basis.

Energy conversion is between different energy forms of different systems that interact with each other, or different energy forms of a single system. However to say energy is "converted" between energy forms is misleading, because energy is not a "thing" that is modified. More precise is to say that "a change in the amount of energy in one energy form leads to an opposite change in another energy form" referring to the origin in single interactions. To understand its basis we need to combine again the first two points above. Conversion is nothing else than the effect that energy appears differently, meaning the ways we observe and take notice of the motion and position of objects and their changes varies, especially if they are individually invisible.

The strength of the concept of energy is describing the effect of interactions between objects in a way that even in complex situations like materials consisting of many particles it is connecting the resulting macroscopic changes. For example, chemical reaction energy, due to changes in the composition of molecules, can be converted to kinetic energy of them, and in bulk then becomes thermal energy which we observe as a change of temperature. And thermal energy can be used to expand and thus deform a gas, related to deformation energy; here we observe the increase in volume. The kinetic energy of an apple, observed by its motion, becomes invisible when the apple hits the ground and comes to rest; it is converted finally to deformation energy and thermal energy of the ground and the apple, both observable only at a close look, or with technical aids. The number of options is endless. The set of energy forms comprises much more options than the fundamental energy contributions (Fig. 5-5, Fig. 5-8). It thus connects an incredible number of different things, for example the volume V and polarization P of a macroscopic object, the number N of the particles of a defined composition a material is made of, and the velocity v of an object and its height h above ground. This is why energy is such a useful and common concept. However, conversion is not arbitrary, it is limited by amount and direction. And not to forget, as discussed in section 2.1.4.3, a connection is needed. The limitation by amount is due to energy conservation. The change of energy in an energy form is balanced by changes in others; their sum is zero. The limitation by direction originates from the fact that overall disorder, and thus also the measure for it, the entropy, is increasing $dS \geq 0$ (Eq. 3-45). Conseqently, $dE_{th} = TdS \geq 0$. There is thus a trend to convert other energy forms to thermal energy, e.g. by friction. The increase of disorder makes the energy less useful for changes with order, e.g. to move an object a distance s against a force F. This also has consequences for work and heat. However, work and heat are energy exchanged between systems, not energy forms. Work or heat can be supplied to a system, and the system can then again supply work or heat (1st law of thermodynamics). We discussed this when looking at thermal / heat engines (section 3.3.3.6), specifically their efficiency (Eq. 3-66). Again energy conservation limits the work done or heat supplied by a system in relation to what is supplied to the system. And the 2nd law, $dS \geq 0$, limits the direction. Specifically, while work can easily be converted to heat completely, the consequence of the 2nd law is that heat cannot be used completely to get work in a cyclic process. Thus work is considered more useful than heat, and this is crucial for our energy system.

5.4.4 Summary and conclusions

Now, at the end of chapter 5, we can see and understand the complete concept of energy, using fundamental energy contributions for the description as well as using energy forms. Actually, we can understand energy forms only if we understand the concept using fundamental energy contributions.

The fundamental energy contributions look at things in detail, specifically the motion and position of objects individually and their mutual interaction. This is where we understand the energy of an object or a system of objects, how energy is exchanged, how energy is converted, and even why energy is conserved. It looks at the details, including the particles in materials, even down to the particles that form the nuclei of atoms, and thus treats things the way often needed by scientists. Moreover, it gives the necessary background to understand the concept of energy the way it is commonly applied, where we do not look at all details, but rather at what is directly observable, usually visually. This is where the concept of energy uses energy forms.

Discussing energy using energy forms relates to what is directly observable. It avoids a description of the motion and interaction of individual microscopic particles in a macroscopic object (or material), thus avoids the complexity of many-particle systems. But as a consequence, discussing energy using the energy forms is also the hardest way to understand what energy is, how it is exchanged, how it is converted, and especially why it is conserved. As we have no information on what goes on in a system like a material it is like a "black box" thus that changes of the internal energy must be described by changes observed macroscopically, for example of the volume, polarization, magnetization, or amount of a material of a given composition. Not being able to see the details also means not being able to understand entropy, which is crucial to understand why the direction of energy conversion is limited. Entropy and its properties in macroscopic thermodynamics can be derived without microscopic details, but not explained and understood. To understand things it is necessary to develop microscopic models that deal with the motion and interaction of the individual microscopic particles. This is done in statistical thermodynamics. Books relating macroscopic observations with the microscopic origin are typically physics and chemistry books (e.g. Atkins 1990, Brown et al. 2015), especially thermodynamics books (e.g. Reif 1985), and books for materials science (de Podesta 2002). A systematic and comprehensive study reveals the basics of energy forms.

Using energy forms the system is always a macroscopic object (or material), and because of this it is fixed what is external and what is internal energy. As a consequence of being able to observe the motion and position of macroscopic objects directly, while not being able to observe the microscopic particles in their interior, external and internal energy have different energy forms (Fig. 5-6, Fig. 5-7). For external energy there is kinetic energy, and potential energy due to the fundamental interactions, which are gravitation, the electrostatic interaction with its relativistic magnetic counterpart, while for internal energy there is thermal, deformation, electric, magnetic, chemical, nuclear, excitation, and ionization energy. The analysis also shows that while for forms of external energy absolute values of energy can be given, for forms of internal energy only small changes, denoted by d, are treated. One reason is that for forms of external energy it is easy to define the zero point (v, ω, E, B are zero), while for internal energy it is not as we only get information on energy changes from observed changes, e.g. of volume V. Another reason is that forms of internal energy can affect each other, e.g. if a solid becomes liquid by a change of its temperature, related to a change of thermal energy, this affects changes of shape, related to changes of deformation energy. And a system can have energy of any of the energy forms, and it can exchange energy by any of the fundamental interactions except the nuclear interactions, and in addition by mechanical and thermal contact. Mechanical and thermal contact are again macroscopic simplifications of complex interactions by particles which are not directly observed.

The energy forms are the way how the concept of energy and what relates to it was discovered in history. The application of the concept started long before it was understood in detail, and this has still significant consequences. The two most fundamental ones are the level of clarity of understanding of the energy forms with their precise definition, and the ability to write them down as a complementary set, needed to make correct energy balances. With regard to understanding of the energy forms and their clear definition we found that common literature is often not specific and precise enough. With regard to the energy forms being a complementary set we found that two energy forms, excitation and ionization, are commonly not mentioned in thermodynamic energy balances. This is not too surprising as excitation and ionization only show up in specific applications. Nevertheless, they are needed to show which set of energy forms is a complementary set, and which are derived effects, like the energy of phase change or of solution.

6 Concluding words

We have now finished our discussion of the basics of energy, and it is time to come to an end. Lets start with a short summary of what we found.

To understand something it is necessary that it is based on familiar things. Thus, we have chosen the common, fundamental concepts about the world around us as stage for the discussion. Space refers to where things are, and time is used to describe when, specifically to describe processes of change. Things we are familiar with are objects, like a stone. Matter refers to what the things are made of; specifically, large objects consist of smaller ones, commonly called particles. Forces relate to why and how things change, specifically the position of objects (including small particles) by motion (change of position). Objects cause interactions, e.g. in a collision or by gravitation, and interactions affect the objects, which is described by forces. These concepts come from everyday life; everybody is familiar with them, and they were thus also the basis of scientific investigations early in history.

Based on this stage are the key ideas for the general strategy of discussion. Macroscopic observations on systems with many objects / particles, like materials, can be understood from their microscopic origin. In a comprehensive treatment, microscopic origins and macroscopic observations must be discussed together. The key idea that we used for the whole discussion is that the origin of energy, its conservation, and its conversion, is in simple situations, which is 1 object in motion, and cases of interaction with a 2nd object. Complex situations are just a combination of many simple situations.

To be able to understand things it is also necessary to stay within the validity of classical mechanics, this means using the concept of forces that we are familiar with from everyday life and Newton's laws of motion, as long as possible. They can still be understood from everyday experience.

Nothing changes if there is no interaction; thus, the basis of everything are interactions. An interaction is always between two objects, e.g. by gravitation or by collision, and it affects both interacting objects. What we are interested in, in simple situations, is ways to describe (explain) the effect of interactions between objects on the motion and position of the objects, or changes thereof (observation), and correspondingly for complex situations.

Throughout history, the concepts of "force", "momentum", and "energy" were developed to describe the effect of interactions between objects.

Interactions act on the interacting objects symmetrically. In the concept of forces this is expressed by Newton's 3rd law. In the concept of momentum, looking at changes with time, it results in the conservation of momentum, and looking at changes of position it results in the conservation of energy. Energy is a concept to describe interactions, not a "thing" that is conserved. Instead of saying that energy is "conserved" we better say that every interaction is associated with an action and a reaction energy of equal magnitude and opposite sign, and consequently overall energy does not change.

If an object falls, the interaction by gravitation with another object results in a change of its velocity. The change of height and the change of velocity are related to the energy forms of gravitational energy and kinetic energy. Thus, due to the interaction both changes that we observe come together, and we call the related change in the energy forms commonly energy conversion.

Thus, from simple situations we can understand why energy is conserved, and also conversion between energy forms. Nothing new happens if many particles interact many times and in different ways in complex situations. As a result, the concept of energy can be understood from a few simple, basic principles and situations, regarding the motion and interaction of objects.

Regarding its physical basis, energy is a concept to describe the effect of interactions between objects on the objects, e.g. their motion and position. The concepts of force and of momentum do this also, each with specific advantages and limitations. In complex situations, because of the complexity we can only understand the basic effects. In materials, where the objects are small particles, we also cannot see the objects, their position or motion. What we are commonly interested in is then changes that we can observe macroscopically, like a change of volume, or of an amount of a substance. These changes are still based on simple situations, but in most cases by complex combinations.

Regarding its use, energy is a concept describing the effect of interactions between objects on the objects motion and position in a way that even in complex situations energy is connecting the resulting macroscopic changes. The terms related to an interaction are a product of a force and a displacement, in complex situations called generalized forces and displacements.

The macroscopic changes resulting from an interaction are quantitatively related to changes in energy forms. For example, the kinetic energy of an object is related to its velocity v, the gravitational energy of an object to its height h, changes of thermal energy to changes of entropy S, and changes of deformation energy to a change of shape, e.g. by a change of volume V.

The concept of energy, based on conservation and conversion, correlates changes in many different things, which is what makes it so useful in science and engineering, and even in everyday life. Energy correlates different energy contributions or forms by "conversion" between them, storage, transport, or energy "exchange". Important for quantitative predictions is that energy is "conserved". It has conservation in common with momentum, however the concept of energy is much more useful as it correlates much more different things, but it is also more complex and harder to understand. Energy is commonly subdivided in different ways: kinetic and potential energy, energy contributions / forms, external and internal energy (thermodynamics), and when energy is exchanged it is subdivided into work and heat. This explains the original statement of energy being the "source" of work, or the ability to do work. We do now understand how work, which we experience directly by a force acting a distance, originates e.g. from people, animals, or combustion heat. It is from interactions on a microscopic scale; we just can't see the origin of it. There is nothing mysterious about it.

The systematic and comprehensive treatment of the concept of energy has also shown more things, besides the physical basis and thus what energy is. It has allowed us to understand the different energy forms, and following, a clear and precise definition of them. This is crucial to understand energy. Specifically, we found that the energy forms of internal energy are significantly different from those of external energy. And it has shown that for a complete set of energy forms, which covers all energy such that none is missing and none counted more than once, two energy forms are usually missing in literature: excitation and ionization energy. By including them, the energy forms become a systematic set that describes the energy related to macroscopic changes. Others, like the energy associated with a phase change from solid to liquid, are combinations of them, often described by enthalpy. Thus, while there are so many different energy forms, and ways how we observe energy, the basis of all is quite simple. This is amazing! And the systematic set of energy forms makes it much easier to understand and apply the full concept, to analyze it, or to explain it.

However, there are also things we cannot explain within the chosen stage. There are things that cannot be described by forces and Newton's laws of motion, thus classical mechanics, some not even by the motion and interaction of objects. But we saw that not all these limitations are as severe as it seems at a first glance. What is not covered by the chosen stage is when forces are not applicable, for example when quantum mechanics applies. For atoms we saw that we can still understand things starting from the quasi-classical model. To some extent this also includes energy of fields, and energy exchange by waves, even quantization by exchange particles. But we cannot explain electromagnetic induction within the chosen stage. That a varying electric field causes a magnetic field, and reverse, obviously means that there is a way that energy is associated with regions of space without presence of matter. And there is more. In an e^- and e^+ annihilation the particles disappear and electromagnetic energy is left, and the reverse in a corresponding creation process. So energy cannot only be associated with regions of space without matter, there is even conversion and conservation between energy and matter. This describes Einstein's formula $E = m \cdot c^2$. But that the concept of objects with mass that interact is a simplification is already clear from the wave nature of matter, so these things just go further. Thus, the general conservation of mass-energy is not that surprising, even if we cannot understand it at all within the stage chosen for discussion here.

Our attempt to understand the basics of energy was thus quite successful. Based on our systematic and comprehensive discussion of the interaction of objects from simple to complex situations, we are now able to understand the basics of energy. Even more, we have been able to put them in a systematic and consistent framework.

I hope you enjoyed reading this book as much as I enjoyed writing it, and that you find it a valuable introduction to the fascinating topic of energy.

Finally, this is the first attempt to cover the basics of energy in a systematic, clear, and comprehensive way, going from simple to complex situations. Despite that a systematic treatment is a way to discover inconsistencies, there is always a chance that a text contains other errors or inaccuracies. Further on, there is always potential to improve things, e.g. explanations or examples. At the end of this book, I thus would like to encourage everybody to give feedback to improve this book, as well as if it was found helpful.

Appendix

Throughout this book, the International System of units (SI) is used.

A1 Prefixes

SI prefixes of decimal numbers

Factor	Prefix	Symbol
10^{24}	yotta	Y
10^{21}	zetta	Z
10^{18}	exa	E
10^{15}	peta	P
10^{12}	tera	T
10^{9}	giga	G
10^{6}	mega	M
10^{3}	kilo	k
10^{2}	hecto	h
10^{1}	deca	da
10^{-1}	deci	d
10^{-2}	centi	c
10^{-3}	milli	m
10^{-6}	micro	µ
10^{-9}	nano	n
10^{-12}	pico	p
10^{-15}	femto	f
10^{-18}	atto	a
10^{-21}	zepto	z
10^{-24}	yocto	y

A2 Units and constants

SI units

quantity		unit		dimension
name	**symbol**	**name**	**symbol**	**symbol**
length	l	Meter	m	L
mass	m	Kilogram	kg	M
time	t	Second	s	T
electric current	I	Ampere	A	I
thermodynamic temperature	T	Kelvin	K	Θ
amount of substance	n	Mole	mol	N
luminous intensity	J	Candela	cd	J

SI supplementary units

quantity		unit	
name	**symbol**	**name**	**symbol**
plane angle	$\alpha, \beta, \phi,$ φ, θ, ψ	radian	rad
solid angle	Ω	steradian	sr

SI derived units

quantity		unit		expressed in	
name	symbol	name	symbol	other SI units	SI base units
general					
area	A				m^2
volume	V				m^3
mass density (density)	ρ				$kg \cdot m^{-3}$
number of particles	N			1	
particle density	n				m^{-3}
frequency	f, ν	Hertz	Hz		s^{-1}
velocity	v				$m \cdot s^{-1}$
angular velocity	ω				$rad \cdot s^{-1}$
acceleration	a				$m \cdot s^{-2}$
angular acceleration					$rad \cdot s^{-2}$
energetic					
energy, work, heat	E, W, Q	Joule	J	N·m	$m^2 \cdot kg \cdot s^{-2}$
power	P	Watt	W	J/s	$m^2 \cdot kg \cdot s^{-3}$
internal energy	U, E_{int}	Joule	J	N·m	$m^2 \cdot kg \cdot s^{-2}$
free energy	F	Joule	J	N·m	$m^2 \cdot kg \cdot s^{-2}$
enthalpy	H	Joule	J	N·m	$m^2 \cdot kg \cdot s^{-2}$
free enthalpy	G	Joule	J	N·m	$m^2 \cdot kg \cdot s^{-2}$
mechanical					
force	F	Newton	N		$m \cdot kg \cdot s^{-2}$
pressure	p	Pascal	Pa	N/m^2	$m^{-1} \cdot kg \cdot s^{-2}$
momentum	p			N·s	$m \cdot kg \cdot s^{-1}$
torque	M			N·m	$m^2 \cdot kg \cdot s^{-2}$
moment of inertia	J, Θ				$kg \cdot m^2$
dynamic viscosity	η, μ	Poise	P	Pa·s	$N \cdot s \cdot m^{-2}$
kinematic viscosity	ν	Stokes	St		$m^2 \cdot s^{-1}$

Continued on next page

quantity		unit		expressed in	
name	symbol	name	symbol	other SI units	SI base units
thermal					
heat capacity	C			J/K	$m^2 \cdot kg \cdot s^{-2} K^{-1}$
specific heat capacity	c				
thermal conductivity	λ			W/(m·K)	$m \cdot kg \cdot s^{-3} K^{-1}$
thermal resistance	R			K/W	$m^{-2} \cdot kg^{-1} \cdot s^3 \cdot K$
entropy	S			J/K	$m^2 \cdot kg \cdot s^{-2} K^{-1}$
electrical					
electric charge	Q	Coulomb	C		$A \cdot s$
electric potential	U	Volt	V	J/C	$m^2 \cdot kg \cdot s^{-3} \cdot A^{-1}$
electric field strength	E			N/C=V/m	$m \cdot kg \cdot s^{-3} \cdot A^{-1}$
electric capacitance	C	Farad	F	C/V	$m^{-2} \cdot kg^{-1} \cdot s^4 \cdot A^2$
permittivity	ε			F/m	$m^{-3} \cdot kg^{-1} \cdot s^4 \cdot A^2$
electric resistance	R	Ohm	Ω	V/A	$m^2 \cdot kg \cdot s^{-3} \cdot A^{-2}$
electric conductivity	σ			$\Omega^{-1} \cdot m^{-1}$	$m^{-3} \cdot kg^{-1} \cdot s^3 \cdot A^2$
electric dipole moment	p			C·m	$A \cdot s \cdot m$
polarization	P			C·m	$A \cdot s \cdot m$
magnetical					
magnetic flux density	B	Tesla	T	Wb/m^2	$kg \cdot s^{-2} \cdot A^{-1}$
magnetic inductance	L	Henry	H	Wb/A	$m^2 \cdot kg \cdot s^{-2} \cdot A^{-2}$
magnetic field strength	H				A/m
permeability	μ			H/m	$m \cdot kg \cdot s^{-2} \cdot A^{-2}$
magnetic dipole moment	μ			N·m^2/Wb	$m^2 \cdot A^{-1}$
magnetization	M			N·m^2/Wb	$m^2 \cdot A^{-1}$
electromagnetical / light					
wavelength	λ	meter	m		m
chemical					
chemical potential	μ			J/mol	$m^2 kg \cdot s^{-2} mol^{-1}$

Non-SI units, which are still in use

quantity		unit		expressed in	
name	**symbol**	**name**	**symbol**	**other SI units**	**SI base units**
volume	V	Liter	l		$10^{-3}\,\mathrm{m}^3$
mass	m	ton	t		$10^3\,\mathrm{kg}$
pressure	P	bar	bar		$10^5\,\mathrm{Pa}$
length	l	Ångström	Å		$10^{-10}\,\mathrm{m}$

Some fundamental natural constants (to three digits accuracy)

quantity		computational value	expressed in	
name	**sym-bol**		**other SI units**	**SI base units**
speed of light in vacuum	c_0	$3.00 \cdot 10^8$		$\mathrm{m \cdot s^{-1}}$
electron charge	e	$1.60 \cdot 10^{-19}$		C
electron mass	m_e	$9.11 \cdot 10^{-31}$		kg
proton mass	m_p	$1.67 \cdot 10^{-27}$		kg
neutron mass	m_n	$1.67 \cdot 10^{-27}$		kg
permittivity constant in vacuum	ε_0	$8.85 \cdot 10^{-12}$	F/m	$\mathrm{A \cdot s \cdot V^{-1} \cdot m^{-2}}$
permeability constant in vacuum	μ_0	$1.26 \cdot 10^{-6}$	H/m	$\mathrm{V \cdot s \cdot A^{-1} \cdot m^{-1}}$
Plank's constant	h	$6.63 \cdot 10^{-34}$	$\mathrm{J \cdot s}$	$\mathrm{m^2 \cdot kg \cdot s^{-1}}$
universal gas constant	R	8.31	$\mathrm{J/(mol \cdot K)}$	$\mathrm{m^2 \cdot kg \cdot s^{-2} \cdot mol^{-1} K^{-1}}$
Avogadro's number	N_A	$6.02 \cdot 10^{23}$	1/mol	$\mathrm{mol^{-1}}$
Boltzmann's constant	k_B	$1.38 \cdot 10^{-23}$	J/K	$\mathrm{m^2 \cdot kg \cdot s^{-2} \cdot K^{-1}}$
Faraday's constant	F	$9.65 \cdot 10^4$	C/mol	$\mathrm{A \cdot s \cdot mol^{-1}}$
Stefan-Boltzmann's constant	σ	$5.67 \cdot 10^{-8}$	$\mathrm{W/(m^2 \cdot K^4)}$	$\mathrm{kg \cdot s^{-3} \cdot K^{-4}}$
gravitational constant	G	$6.67 \cdot 10^{-11}$		$\mathrm{N \cdot m^2 \cdot kg^{-2}}$

A3 Nomenclature

Superscripts

superscript	meaning
+, -	electric charge
*	excited

Subscripts

subscript	meaning	subscript	meaning
sys	system	sur	surrounding
int	internal	ext	external
kin	kinetic	pot	potential
trans	translational	r	relative
rot	rotational	tot	total
g	gravitation(al)	rad	radiation
el	electric	e	electron
mag	magnetic	n	neutron
def	deformation	p	proton
th	thermal	he	heat engine
chem	chemical	ideal	ideal
nuc	nuclear	p	at constant pressure
ex	excitation	V	at constant volume
ion	ionization	i, j, n, m, x	integer numbers
mech	mechanical	1, 2, …	integer numbers
hand	hand	0	vacuum
in	in	i	initial
out	out	f	final
A	Avogadro	c	centrifugal
B	Boltzmann	f	friction
b	binding		

Abbreviations (acronyms)

abbreviation	meaning
SI	International System of units
QM	Quantum Mechanics

Physical and chemical nomenclature for elements and compounds

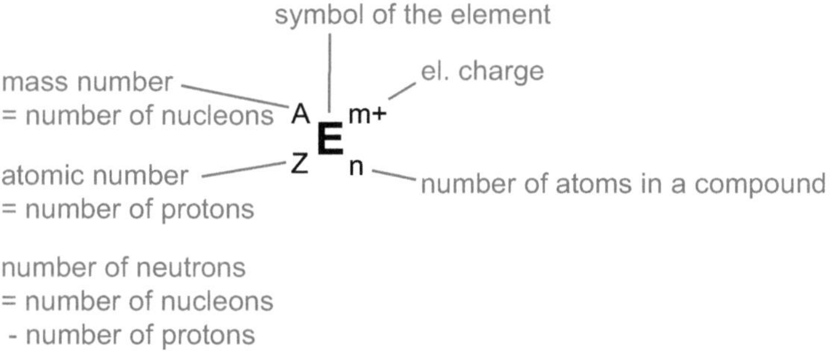

References

Atkins 1990 P.W. Atkins: Physical chemistry. Oxford University Press, 4th ed., 1990, ISBN 0-19-855293-1

Benson 1991 H. Benson: University physics. John Wiley & Sons Inc., 1991, ISBN: 0-471-60528-X

Bransden and Joachain 1983 B.H. Bransden and C.J. Joachain: Physics of atoms and molecules. Wiley, 1983, ISBN 0-582-44401-2

Brown et al. 2015 T.L. Brown, H.E. LeMay, B.E. Bursten, C.J. Murphy, P.M. Woodward, and M.W. Stoltzfus: Chemistry – The central science. Pearson, 13th ed., 2015, ISBN-13: 978-1-292-05771-2

Çengel 1998 Y.A. Çengel: Heat transfer – A practical approach. McGraw-Hill, 1998, ISBN 0-07-115223-7

Çengel and Boles 2002 Y.A. Çengel, and M.A. Boles: Thermodynamics – An engineering approach. McGraw-Hill, 4th ed., 2002, ISBN 0-07-112177-3

Cleveland and Morris 2009 C.J. Cleveland and C. Morris (eds.): Dictionary of energy. Elsevier, 2009, ISBN 978-0-08-096491-1

Dawson-Tunik 2004 Dawson-Tunik: It Has Bounciness Inside! Developing Conceptions of Energy, 2004, *http://devtestservice.org/PDF/Bounciness.pdf*

de Podesta 2002 M. de Podesta: Understanding the properties of matter. Taylor & Francis, 2nd ed. 2002, ISBN 0-415-25788-3

English Oxford living dictionaries 2017 English Oxford living dictionaries: Energy. *www.en.oxforddictionaries.com/definition/Energy*, accessed July 27th, 2017

Faires and Simmang 1978 V.M. Faires and C.M. Simmang: Thermodynamics. 6th ed., Macmillan Publishing Co., Inc., 1978, ISBN 0-02-335530-1 (hardbound), ISBN 0-02-978910-9 (international edition)

Fermi 1956 E. Fermi: Thermodynamics. Dover, 1956, ISBN 978-0-48-660361-2 (originally published by Prentice-Hall, 1937)

Gaskell and Laughlin 2018 D.R. Gaskell and D.E. Laughlin: Introduction to the Thermodynamics of Materials. CRC Press, 6th ed., 2018, ISBN 978-1-4987-5700-3 (hardback)

Green 2006 B. Green: Das elegante Universum. Goldmann, 2006, ISBN: 978-3-442-15374-9 (originally published as: The elegant universe. Superstrings, hidden dimensions, and the quest for the ultimate theory. Norton Company, 1999)

Halliday et al. 1993 D. Halliday, R. Resnik and J. Walker: Fundamentals of Physics. John Wiley & Sons, 4th ed., 1993, ISBN 0-471-59986-7

Hari Dass 2014 N.D. Hari Dass: The Principles of Thermodynamics. CRC Press, 2014, ISBN 978-1-4665-1208-5 (hardback)

Hornby 1983 A.S. Hornby: Oxford advanced learner's dictionary of current English. Oxford University press, 14th edition, 1983, ISBN 3-8109-0048-6

Kube 2009 J. Kube: Physik - Ein Schnellkurs. Dumont, 2009, ISBN 978-3-8321-9142-9

Kurzweil et al. 2009 P. Kurzweil, B. Frenzel, and F. Gebhard: Physik Formelsammlung. Vieweg + Teubner, Wiesbaden, 2nd ed., 2009, ISBN 978-3-8348-0875-2

learner.org 2017 Physics for the 21st century. Unit 2: The Fundamental Interactions. *https://www.learner.org/courses/physics/unit/pdfs/unit2.pdf*, accessed July 25th, 2017

McPhee 2015 I. McPhee: Physik – Von der Schwerkraft bis zur Quantenmechanik. Librero, 2015, ISBN 978-90-8998-487-6. (english original: The bedside book of physics. Quid Publishing, 2011)

Mehling 2016 H. Mehling: Technologies of energy conversion, storage, and transport in the energy system - A brief introduction. Tredition, 2016, ISBN: 978-3-7345-4049-3

Michelini and Stefanel 2010 M. Michelini and A. Stefanel: Approaches and learning problems in energy teaching/learning: an overview. Contribution to the report of the workshop: Teaching about energy: Which concepts should be taught at which educational level? Reims 2010 *http://www.fisica.uniud.it/URDF/articoli/ftp/2011/2011-13-14.pdf*

Morse 1969 P.M. Morse: Thermal Physics. The Benjamin / Cummings Publishing Company Inc., 1969, ISBN 0-8053-7202-4

Nicholls and Ogborn 1993 G. Nicholls, and J. Ogborn: Dimensions of children's conceptions of energy. IJSE, 15(1), 73-81(1993)

Quaschning 2005 V. Quaschning: Understanding renewable energy systems. Earthscan, 2005, ISBN 1-84407-128-6

Reif 1985 F. Reif; Fundamentals of statistical and thermal physics. McGraw-Hill International editions, 1985, ISBN 0-07-085615-X, 1985

Rooney 2011 A. Rooney: The story of physics. Arcturus Publishing Limited, 2011, ISBN 978-1-84837-769-1

Sears and Zemansky 2016 Sears and Zemansky: University Physics with Modern Physics. Pearson, Global edition, 14th ed., 2016, ISBN 978-1-292-10031-9

Smil 1999 V. Smil: Energies - An Illustrated Guide to the Biosphere and Civilization. The MIT Press, Cambridge Massachusetts & London, England, 1999, ISBN 0-262-19410-4

Snurr and Freude 2014 R. Snurr and D. Freude: Energy definition; Energy Fundamentals. *http://www.energy-fundamentals.eu/01.html*, last changed 01.11.2014, accessed 28.11.2016

Störig 2004 H.J. Störig: Kleine Geschichte der Wissenschaft. Part 1. Parkland Verlag, 2004, ISBN: 3-89340-056-7

Tipler and Mosca 2004 P.A. Tipler and G. Mosca: Physics for scientists and engineers – Volume 1 Mechanics, oscillations, and waves, thermodynamics. Freeman and Company, 2004, ISBN: 0-7167-0809-4

Trumper 1993 R. Trumper: Children's energy concepts: A cross-age study. International Journal of Science Education, 15, 139 - 148. (1993).

Watson 2014 D. Watson: FT Exploring web site, section: What is the Definition of Energy? *http://www.ftexploring.com/energy/definition.html*, accessed Nov. 2nd 2016

Walker et al. 2014 J. Walker, D. Halliday, and R. Resnick: Principles of physics. 10th ed., international student version, Wiley, 2014, ISBN: 978-1-118-23074-9

Wikipedia 2016a Wikipedia: Fundamental interaction. *https://en.wikipedia.org/w/index.php?title=Fundamental_interaction&oldid=724570704*, accessed July 7th 2016

Wikipedia 2016b Wikipedia: Conservation of energy. 2016,

Wikipedia 2017a Wikipedia: Work (physics). 2017, *https://en.wikipedia.org/wiki/Work_(physics)?oldid=777048201*

Wikipedia 2017b Wikipedia: Nuclear force. *https://en.wikipedia.org/w/index.php?title=Nuclear_force&oldid=787363859*, acessed June 29th 2017

Index